AF801814

An Archaeology and History of a Caribbean Sugar Plantation on Antigua

Florida Museum of Natural History: Ripley P. Bullen Series

An Archaeology and History of a Caribbean Sugar Plantation on Antigua

EDITED BY GEORGIA L. FOX

University of Florida Press
Gainesville

Copyright 2020 by Georgia L. Fox
All rights reserved
Published in the United States of America.

25 24 23 22 21 20 6 5 4 3 2 1

Library of Congress Cataloging-in-Publication Data
Names: Fox, Georgia Lynne, 1954– editor.
Title: An archaeology and history of a Caribbean sugar plantation on Antigua / Georgia L. Fox.
Other titles: Ripley P. Bullen series.
Description: Gainesville : University of Florida Press, 2020. | Series: Florida Museum of Natural History: Ripley P. Bullen series | Includes bibliographical references and index.
Identifiers: LCCN 2019031038 (print) | LCCN 2019031039 (ebook) | ISBN 9781683401285 (hardback) | ISBN 9781683401445 (pdf)
Subjects: LCSH: Sugar trade—Antigua and Barbuda—History. | Plantations—Antigua and Barbuda—History. | Betty's Hope (Antigua and Barbuda)—History. | Betty's Hope (Antigua and Barbuda)—Antiquities.
Classification: LCC HD9114.A632 A63 2020 (print) | LCC HD9114.A632 (ebook) | DDC 338.7/633610972974—dc23
LC record available at https://lccn.loc.gov/2019031038
LC ebook record available at https://lccn.loc.gov/2019031039

UF PRESS
UNIVERSITY OF FLORIDA

University of Florida Press
2046 NE Waldo Road
Suite 2100
Gainesville, FL 32609
http://upress.ufl.edu

CONTENTS

FIGURES

TABLES

ACKNOWLEDGMENTS

Archaeologists are used to working in teams, relying on the combined skills and expertise of all involved. The same can be said for Betty's Hope. First and foremost, the project would not have been possible without the support and guidance of Dr. Reginald Murphy, who wears several hats as Antigua and Barbuda's archaeologist, as director of heritage resources, research, archaeology, and environment at Nelson's Dockyard National Park, and as secretary general for the National Commission for UNESCO for Antigua and Barbuda. Reg's unfailing sense of humor, razor-sharp intellect, and encyclopedic knowledge of Caribbean archaeology and prehistory has been a joy. We are also deeply grateful to the government of Antigua and Barbuda, the Betty's Hope Trust, the Antigua and Barbuda Historical and Archaeological Society, and Michele Henry, director of the Museum of Antigua and Barbuda, for their ongoing support. I am eternally grateful to Mr. Joseph Prosper, director of the National Archives of Antigua and Barbuda, for allowing me access to the original Codrington Papers, providing the opportunity to discover critical details in the documents that only that type of access can provide. I would also like to thank Agnes Meeker for sharing her extensive knowledge about Antigua's plantation history and introducing me to many of Antigua's old plantation sites.

Directing an archaeological field school for 10 years involves a number of institutions and individuals to thank. I am grateful to president Gayle Hutchinson of California State University, Chico, dean Eddie Vela of the College of Behavioral and Social Sciences, the Department of Anthropology at California State University, Chico, as well as the staff at the Regional Continuing Education Office. I would also like to thank the Office of International Education, the Office of Risk Management, and the University Research Foundation. I am grateful to James Scolero, who was integral in the early days of the project. It was during our first survey that we discovered that the buildings at Betty's Hope aligned to true north. I also would like to thank my colleagues in the Department of Anthropology, who were supportive of my efforts to develop the field school and project.

Generous funding from the David W. and Helen E. F. Lantis Fellowship allowed our research team to expand, notably by the addition of Christian Wells, whose expertise in soils and environmental archaeology was fundamental to the project. Over the years, a number of teaching assistants were integral in the development of the field school, including Melinda Button, Leanna Flaherty, Katrina Eichner, Maija Glasier-Lawson, and Charlotte Goudge. I would like to thank Cory Look for his long hours of survey work, map making, and teaching students how to level a total station. I am especially grateful to Christopher Waters, who started out as a field school student and subsequently served as my teaching assistant for several years. His tireless good cheer, endless trips to town, and transporting of students to and from the site kept us going. No acknowledgment would be complete, however, without thanking all of the over 100 wonderful field school students who participated in the project, some of whom returned for more than one field season, enduring the heat and fire ants, before going on to professional careers in archaeology and museum studies. Without them, the project would not have been possible, and I am grateful to them for their hard work and inspiration.

Special thanks are due to every contributor to this volume, bringing their expertise, ideas, and collegiality to the project. Attending the annual meeting for the Association of Caribbean Historians in 2013 resulted in rich intellectual collaborations and friendships, reinforcing collaboration between historians and anthropologists. Gratitude is also extended to my colleagues in Caribbean archaeology for their friendship and valuable insights. Their work is exceptional, and I feel fortunate to be part of this cohort. I wish to extend my gratitude to the peer-reviewers of this volume, whose comments and suggestions provided the necessary guidance in improving the manuscript.

Finally, I would like to thank the Bardoe and Murphy families for their kind and generous support, and Tamara Varney for her encouragement to pursue research on Antigua. Most of all, during times of pure exhaustion and long hours, Nicki Murphy fortified us with her magnificent cooking, good humor, and eternal wisdom, without which we could not have done this project. Thank you, Nicki! We are especially grateful to the people of Antigua for their ongoing kindness and faith in our efforts to better understand the shared history and legacy of their ancestors and all they endured and inspired in the generations that followed and for those to come.

FOREWORD

Plantations are uniquely interesting places. In the post-Columbian world, they were centerpieces of colonialist activity, being some of the first places where peoples of European and African ancestry interacted on a daily basis. This single characteristic bestows upon plantations a centrally important role in world history. The colonialist intent of early plantations rightfully attracted the attention of many documentary historians at an early date. Prominent members of European and American society owned the largest estates, and they were the sort of individuals who wrote about themselves and were written about by others. It took some time for historians as a group to turn their gaze away from these often-revered men, but when they did, they made prodigious strides in plugging the voids in knowledge about the daily lives of all the other, much more numerous individuals who lived at the plantation. With time, historians began to look beyond the walls of the plantation mansions and into the industrial buildings and homes of the enslaved men and women who made plantations successful enterprises. The efforts of historians, as positive as they frequently were, however, were nonetheless woefully incomplete. Not until archaeologists began to dig into the earth were the material realities of past plantation life made tangible, visible, and real. Dedicated archaeologists began to construct new foundations of cultural and historical understanding as each new excavation added new bricks to the edifice of knowledge. Each new excavation revealed something heretofore unimagined: the presence of firearms in the homes of the enslaved, the remarkable range of wild and domestic foods eaten by plantation residents, and the conscious adjustment of landscapes to reflect the ongoing tensions between social relations and plantation production. Historians could never have seen these often-subtle clues because such traces were buried in the earth and usually unmentioned in the frequently scant historical record. Contemporary observers seldom thought to memorialize the commonplace in their writings, perhaps believing them simply too irrelevant to mention.

The efforts of archaeologists to investigate plantation life did not come easy. For too many years, historical archaeologists, like their historian forebears, concentrated only on big houses. When I had my first hands-on experience with plantation archaeology in 1973, I found it curious that the chief archaeologist had little interest beyond the plantation's mansion and its immediate yard. The houses of the men, women, and children who were held in bondage were not in the project's purview, even though Charles Fairbanks had already demonstrated the many promises contained within the archaeology of slave life. Historical archaeology at the time was still struggling to find its voice, so perhaps we should not be surprised that some excavators adopted the conservative mindset of early plantation archaeology and chose simply to follow the lead of others.

Archaeologists who adopted an explicitly anthropological approach, however, soon turned their focus onto enslavement, a step mirroring the creation of social, or bottom-up, history. A concern with enslavement necessarily meant an interest in colonial and antebellum estates, and about a decade would pass before historical archaeologists would develop an equal interest in the transition from bondage to freedom. This transition on plantations was materially expressed in the homes of tenant farmers, in the use of plantation space, and in the character of the material culture. This transformation of focus in turn led to an interest in the role of racial assignment and racism on postbellum plantations, while also providing a fresh perspective on racial categorization on antebellum estates. One outcome of this enlightened view was to shine a harsh light on the legacy of the plantation-as-institution. Plantations as economic powerhouses may have largely diminished in power along with the end of state-sponsored enforced bondage, but many of the plantation's most hurtful features continued to exert pressure on daily life well into the twentieth century and arguably even into the present in some places.

In recent years, the archaeological examination of plantation life has become sophisticated and nuanced. This volume exemplifies this maturation as it points to exciting new avenues of research. The Betty's Hope project substantiates the profound significance of continued archaeological research at plantation sites, and with its vitality, it exemplifies the interpretive power of historical archaeology when applied with sensitivity and care.

At Betty's Hope we witness the successful union of meticulous archaeological research with a deep concern for the continuing life of the plantation. The authors take us into the history of the site and introduce its inhabitants with all the rigor of anthropological historical archaeology.

The plantation is not simply a place of the dead. The landscape is as alive as ever, though the inhabitants are now tech-savvy archaeologists, preservationists, and historians, in addition to a vibrant and engaged local community. Some plantation residents speak through the remarkable 300 years of historical documentation while others make their presence known through the physical remains left behind. In the best tradition of anthropological historical archaeology, we learn to appreciate the significance of the estate on the global stage as a producer of one of the modern world's most desired commodities and discover the most intimate details of the residents' material culture. We come to understand in turn that the Betty's Hope research team fully appreciates and respects the continuing importance of this old plantation—past, present, and future.

The Betty's Hope project is a great success. The research represented here provides an intriguing peek into what the future may hold as the excavations continue.

Charles E. Orser Jr.

PART I

Introduction to Betty's Hope Plantation

1

Introduction

GEORGIA L. FOX

The advertisement quietly appeared in the *West India Committee Circular* of April 1942. An old sugar plantation and its surrounding properties were up for sale on the Caribbean island of Antigua, now a British outpost in a region that shouted "backwater," a land characterized by pith-helmeted colonials, decrepit old buildings, and seemingly pliant locals. The ad touted a sugar plantation "containing about 1,400 acres as going concern. Good buildings, growing crops, live and dead stock, appliances, etc." (NAAB Codrington Papers:C67). The problem was that no one wanted it.

Letters flew back and forth between managers and lawyers. The owner, comfortably ensconced in his palatial Gloucester estate at Dodington, had high hopes that were dashed again and again. This rollercoaster continued for some time, even as far back as 1927—as the thick bundles of letters in the archives testify—chronicling the mounting frustrations of Sir Christopher William G. H. Codrington, who was ready to be done with this albatross of a property. Relief finally came in 1944, as the Betty's Hope Plantation and adjacent Codrington-owned lands were finally unloaded by the British-owned Antigua Syndicate Estates, Ltd. A legacy of almost 300 years had finally ended.

Since the publication of Sidney Mintz's powerful and seminal book, *Sweetness and Power*, in 1985, the topics of sugar plantations and slavery have grown and expanded through the collective research of historians, anthropologists, archaeologists, geographers, and other scholars. This book addresses these topics by taking a holistic anthropological approach to the study of one British Caribbean plantation, Betty's Hope, located on the Caribbean island of Antigua, through multiple lines of research as demonstrated in this edited volume.

Although Barry Higman (1984:3) rightly cautioned against focusing on one plantation as representative of a society at large, individual plantation studies can add to the comparative body of knowledge in Caribbean plantation research and how the plantations fit within the broader picture of colonialism and its aftermath in the Americas and the Atlantic World. New avenues of enquiry, approaches in archaeological field methodology, and a greater comparative body of data from which to draw have all contributed to an increased understanding of the sugar plantation and its many facets. By taking a multifaceted approach to the study of one plantation site, we hope to add our insights and research to the existing literature, building on new directions and ideas about a critical period in history that changed the direction of global powers, influences, and economy, as well as the human consumption of commodities that revolutionized and instigated new patterns of behavior.

Although Betty's Hope's history can be generally divided into pre- and post-Emancipation periods, its phases are not always clear cut; the plantation existed with the context of cultural continuums, operating in a dynamic cultural milieu in which various actors adapted and changed, forged new identities, and often resisted imposed structures, long after August 1, 1834, the day of the liberation of enslaved Africans. Time and again, the story of Betty's Hope and places like it are stories of dichotomies—of pain and joy, despair and hope. Above all, they are stories of resilience and survival in the face of the crushing realities of slavery.

Theoretical Perspectives

A number of theoretical frameworks have been applied to our work at Betty's Hope. One that has been particularly useful is Ian Hodder's entanglement theory, derived from quantum mechanics and adopted by social scientists in explaining networks and their inherent dependencies. As societies evolve and change, they are not isolated, orderly, cohesive things, but rather operate in more chaotic, open-ended processes that are embroiled in uncertainty and contingency (Hodder 2012:5). Dependency therefore creates a dialectical struggle within the various entanglements. In the case of Antigua and the British sugar plantocracy, the entire system of animals, plants, and people was based on an accepted premise of economic success and the deep dependency on enslaved labor, making for acute vulnerabilities in the system as a whole.

The entangled mess of cane agriculture in all its manifestations exacted a huge toll in resources and human lives. Overreliance on an untenable lifestyle, combined with the dependency on water availability, unpredictable markets, and armed conflict, created a collective tipping point. If one link in the system failed, others would follow. The archaeological and documentary evidence suggest that in the long term, the system was unsustainable, leading to eventual collapse as the sugar market bottomed out and formerly enslaved peoples found new freedom.

Archaeology at Betty's Hope

The road leading up to Betty's Hope today is striking; two iconic windmills, one of which has been painstakingly restored, greet visitors on their way to the former plantation. The restored mill has served as the backdrop for innumerable field school group photos and is the safe haven in which to duck for cover during the intermittent summer rain showers. The site is located in the central-eastern part of Antigua, which is itself located in the Leeward Islands (figure 1.1).

The Betty's Hope Archaeological Project officially began in the summer of 2007 as a small field school consisting of seven students. The project has since grown to include a wide array of researchers. Working at Betty's Hope has been instructive, surprising, and always rewarding. The project began after an initial visit to Antigua in 2004 to meet with Reginald Murphy, Antigua and Barbuda's chief archaeologist and director of the Nelson's Dockyard Museum, which is now part of a UNESCO World Heritage site. A return trip in 2005 to locate potential sites followed, with the official establishment of the project in 2007. The project continues, but after nine years of archaeological investigations, it is time to share our findings to showcase the variety and extent of the work being conducted at the site and in related projects on Antigua.

From the outset, the goal of the project was to create multiple lines of research that could be integrated into a more holistic study, situated within the Caribbean region and the broader context of the Atlantic World. Work at Betty's Hope has not been without its challenges. After abandonment of the site, vandalism and theft of carved building stone have left gaps where formidable structures once stood. Dense thickets of thorny acacia have taken over the site, and climbing vines grow on the old still house. Site deterioration has been exacerbated by both extreme weather events and

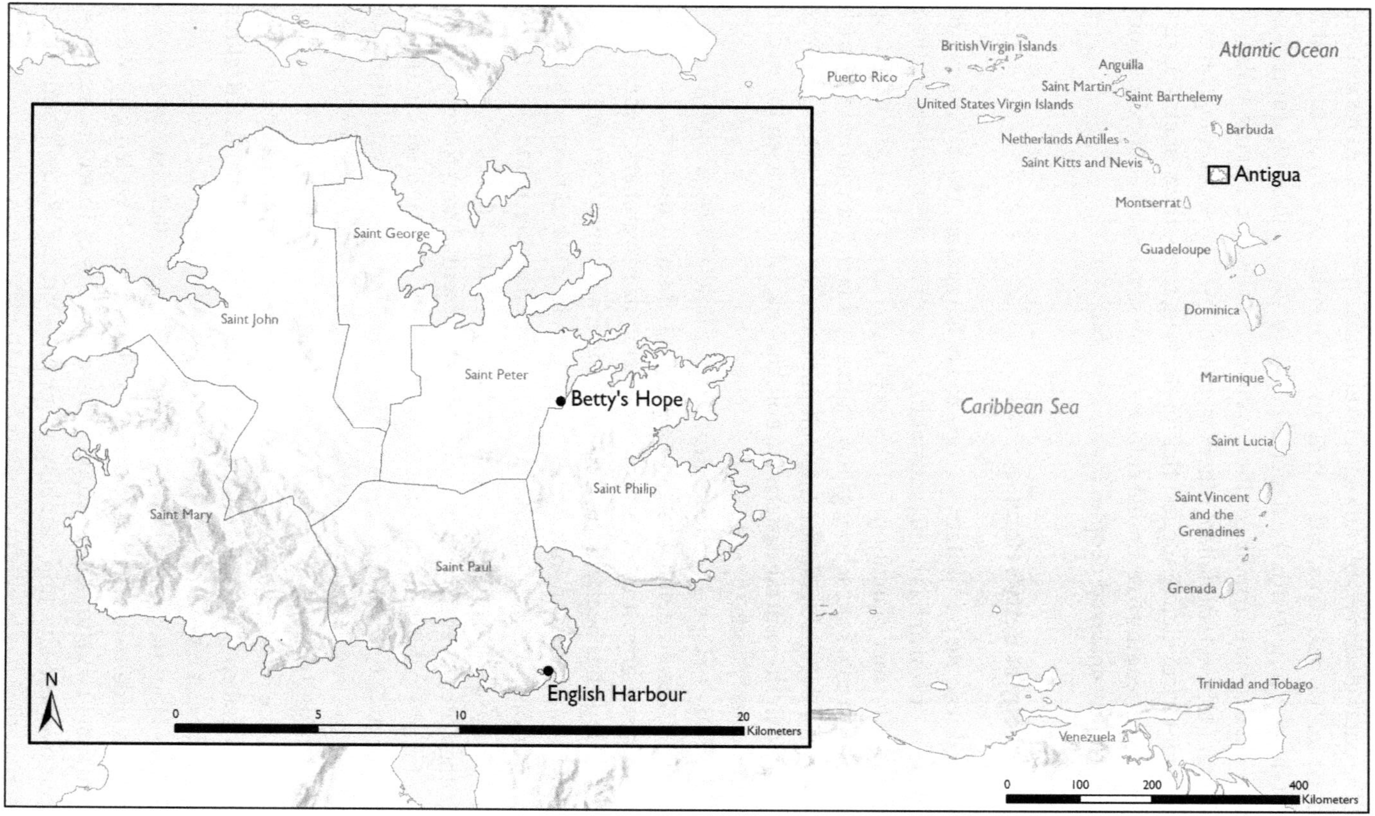

Figure 1.1. Location of Antigua and Betty's Hope. Created by Peter Hansen.

free-roaming herds of sheep and goats—and more recently, donkeys—and by the errant hog and occasional cow. The vegetation and the uneven ground prohibit traditional remote sensing techniques involving ground-penetrating radar, but drone technology has been helpful in identifying hidden structures. The greatest impact, however, has been from human activity, which in recent years has included a group of young tourists who thoughtlessly race their rental ATVs around the historic property.

The site has seen worse, however. The enormity of the site's dark past is never lost on us as we delve into its secrets. Situated on 50 acres and under the aegis of the Betty's Hope Trust, the site comprises a number of extant buildings in various states of disrepair, including the still house, the two windmills, and four deep cisterns. Ruins of the former manager's house, which burned down in 1986, are still visible among the other ruins scattered throughout the property. A small restored visitors center, once the site of a former stable, provides small exhibits for the site's frequent year-round visitors, and interpretive signage dots the landscape near key landmarks like the still house. The boiling house is long gone, as is the great house, which served as the seat of the English Leeward Islands government under Christopher Codrington II and III (see chapter 2).

It is here that the first five years of excavations began. Employing 2 m^2 excavation units, eighteenth-century stone foundations were first revealed at the end of the 2007 field season. For the next nine years, numerous pedestrian surveys were conducted utilizing total stations and GPS units and applying georeferencing utilizing a geographic information system with satellite imagery and old maps, as well as photogrammetry via drone technology. Excavations of the great house and adjoining kitchen took place from 2007 to 2012. A portion of the still house was excavated beginning in 2013, and from 2014 to 2015, excavations were conducted by Charlotte Goudge as part of her doctoral research. Beginning in 2014, the long-awaited excavations began in the area of the slave quarters, previously hampered by the uncertainty of its location. Work continued in the heart of this area in 2015. The 2014–2016 field seasons comprised more surveys and soil sampling, the results of which are discussed in this volume.

One of the challenges of working at Betty's Hope is grasping an understanding of its stratigraphy. Because of the long-term occupation of the site, well into the early 1970s, there exists site disturbance mostly from fill, particularly in the still house area. Yet natural stratigraphic profiles have been clearly identified and delineated, and the artifact assemblages clearly represent all three excavated areas of the site, even with the fill in the still house area. The reuse of carved limestone blocks has also posed some interpretive challenges. Yet building additions and repairs have been identified, particularly in the survey work of Geneviève Godbout, as discussed in chapter 5.

Although the physical challenges presented at the site are mostly surmountable, another aspect of the project—archival research—has not been without its own trials. Until recently, access to the original Codrington Papers in the National Archives of Antigua and Barbuda had been closed off to researchers. The saga of the Codrington Papers has been published elsewhere (Barber 2012). In place of consulting the original papers, researchers utilized microfilm copies that revealed their own limitations. This was made especially apparent when permission was finally granted to study the original papers in the spring of 2016.

Whole sections of the 1710 and 1755 maps that have been critical to our research had either been cut off or were not included in the microfilm versions. Discovering features on the 1710 map in particular was both revelatory and surprising, as multiple versions of the original maps were discovered in almost pristine condition. One of the most important features on the original 1710 map was the location of an early kiln site for the Codrington estates, which we were able to survey after examining the map. One of the reasons Betty's Hope was chosen for long-term research was the copious archive comprising almost 300 years of correspondence, accounts, estate papers, and other materials related to the Codrington landholdings in Antigua and Barbuda.

Relevancy and Contribution

British sugar plantations are complex entities, worlds unto themselves, and they have been dedicated places of study because they "represent a microcosm of broader society" (Wilkie 2004:110). As landscapes deeply imbued with multiple meanings, plantations offer researchers opportunities to view a deeply intertwined world of contested power relations, slavery, environmental change, and a variety of human endeavors occurring over a span of several hundred years (Wilkie 2004:111). Twenty-first-century historical archaeology is moving in exciting directions to tackle commonly held assumptions about plantation life, such as the domination/

resistance model so characteristically associated with British colonialism (see Lyons and Papadopoulos 2002).

Equipped to compare the rich archival resources to the archaeological record and to ethno-historical sources, researchers can illuminate the varied tapestry of the plantation landscape as seen from a variety of different multiscalar approaches and perspectives (Orser 2010). Plantations are dynamic spaces, and plantation studies have evolved into areas of research that go beyond traditional studies of slavery and sugar production. Research in marginal spaces and post-emancipation dynamics, as well as important West African connections, are some of the interesting and thoughtful new directions taking place in plantation studies (Armstrong 2003; Kelly and Ibrahima 2015; Lightfoot 2015; Reilly 2014). Building on the shoulders of giants (Hornborg and Crumley 2007; Pulsipher 1986; Watts 1987), emerging studies on the impacts of deforestation and land-use practices are promising avenues for research (Neiman 2008, 2018). Scientific field sampling and contemporary ethnographic fieldwork chronicle changing landscapes over time from the microscope to on-the-ground informant interviews with local farmers, as demonstrated in this volume and elsewhere (Wells et al. 2017; Wells et al. 2018a; Wells et al. 2018b).

It has been argued that plantation agriculture symbolizes the nexus of a nascent capitalist world system and the creation of the British Empire (Menard 2007:312; Mintz 1974, 1985). Crops like sugar and tobacco were predicated on consumer demand, making the Caribbean an ideal template for the study of developments in labor and capital for the early modern era (see Fox 2015). Formative influences such as the burgeoning Industrial Revolution of the eighteenth century, the Atlantic slave trade, and British colonial policies connected and affected all those involved, where "political and ideological shock waves . . . passed backward and forward between the centers of a world which was already linked" (Bayly 2004:41).

Yet the study of the sugar plantation is more than one of economy. It is also a landscape of shared histories of dependent relationships, violence, subjugation, deepening inequalities, and large-scale destruction of communities and ecosystems. Caribbean plantation studies can therefore provide a window into the tensions and dynamics that occurred within the period of early modernization (Beckles 1997; Delle 2014; Hall 2002; Mintz 1974; Richardson 1987; Stoler and Cooper 1997; Trouillot 1992,

2004). Drawing on Hauser and Hicks (2007:251), "Landscape archaeologies and postcolonial archaeologies face a common challenge: the acknowledgment of material conditions and multiple geographical scales as well as purely ideational accounts of landscape or colonialism." In that vein, embarking on the study of a single site in this volume, we consider the materiality of colonialism at Betty's Hope in many of its various scales of physical and mental spaces as an opportunity to develop better understandings of this important chapter in human history, one that resonates again and again against the backdrop of an ever-changing world that still faces some of the same underlying and troubling issues.

In fact, the recent emphasis on the intertwined nature of political, economic, and social factors with environmental issues and long-term change renders the archaeology of plantations an excellent fit for theoretical frameworks in political ecology as we face the realities of climate change and the politically motivated responses to it. Because of their experience in studying the relational nature of material processes and places within multiple time scales, archaeologists are uniquely positioned to address issues of land use, inequality, and postcolonial legacies (see Ashmore 2018; Morehart et al. 2018).

Antigua as the Setting

Human presence on Antigua dates to at least 2,000 years ago (Murphy 1999:265, 313). Unlike other Caribbean islands, at present, there is no archaeological evidence to support an Amerindian presence at the time of the first recorded arrival of English settlers in 1632. Prehistoric and historical archaeology are rich areas for archaeological investigations on Antigua and Barbuda, with scientific archaeological excavations being conducted on the island for the last 25 years. Most historical archaeology on Antigua has focused on military forts, with plantation archaeology being a recent development. Current maritime archaeology is still in its preliminary stages, as full-scale excavation is hampered by the lack of a suitable full-time conservation facility. Antigua's greatest threat to archaeological sites derives from active land development projects, usually geared toward tourism, the island's economic engine since independence in 1981.

At present, no other sugar plantation had been scientifically excavated on Antigua. The first systematic excavation at Betty's Hope was conducted by Edith Gonzalez in January of 2000 as part of her doctoral research at

Brooklyn College. In 2007, an archaeological field school was established through the Department of Anthropology at California State University, Chico. Reginald Murphy, director of heritage resources for the National Parks Antigua, is chairman of the Betty's Hope Estate Project, and I have served as principal investigator of the project.

The project has had local support from the Betty's Hope Trust and the Museum of Antigua and Barbuda, as well as through educational exchanges and local media exposure. Over the years, tour groups, public presentations at the Museum of Antigua and Barbuda, newspaper articles, radio talk shows, television programs, and print media have informed Antigua's public about our archaeological research. Local input has been critical to and integral in our research design, and the project has encompassed not only archaeology, but also archaeological conservation, museum studies, museum collections care, video production, and cultural heritage management as a holistic approach, as reflected in this volume.

The special circumstances of working in a hot and humid climate that is also prone to the risks of hurricanes and earthquakes present interesting challenges for archaeological curation on the island. This, along with public education and outreach, has informed our active engagement in consulting with Antiguans, working with locals in creating exhibitions in the island's three museums and engaging in other related activities, such as the creation of a small temporary conservation facility at Nelson's Dockyard. We could not have done our work without the generous support of Antiguans.

The material culture recovered from Betty's Hope is curated at the Field Research Station located at Shirley Heights, under the auspices of Antigua's National Parks Authority. Museum studies graduate students in California State University's Department of Anthropology Museum Studies Program have been involved in creating exhibitions at the Museum of Antigua and Barbuda, the Dockyard Museum at Nelson's Dockyard, and the Betty's Hope Visitors Center. In 2012, we produced an educational video on the archaeology of Betty's Hope as a student-led project through the department's Advanced Laboratory of Visual Anthropology.

Research questions guiding our investigations are as numerous and varied as the authors in this volume, although underlying the overall archaeological inquiry is the desire to achieve a better understanding of daily life on a sugar plantation. Our most recent research has concentrated on the long-term impacts of cane agriculture, and recent efforts are being directed toward reconstruction of a pre-Contact environment to develop a more deep-time profile of the island's natural history and human occupation. Betty's Hope was fortuitously imposed on good land. Located in Antigua's fertile Central Plain, the plantation's founder, governor Christopher Keynell, could not have found a better place to cultivate sugarcane. The nearby coastline provided workable anchorages and wharfages for loading and unloading cargo, and the weather could be decent for a good part of the year, with the exception of drought, hurricanes, and storms.

The origin of the name Betty's Hope remains uncertain. The name has often been ascribed to a daughter in the Codrington family, a mistake that is often repeated in the literature and on the internet. Records in the British National Archives, however, indicate that the name was well in place during Christopher Keynell's tenure (NACO:Acts of Antigua), making it more likely that the estate was named after Joan Elizabeth Keynell, who went by "Betty" (Barber 2012:3; Dyde 2000:295; Nicholson and Dyde 2014:11; Oliver 1894–1899:II:123). According to Dyde (2000:295), the "Hope" part of the plantation's name refers to an antiquated term used to describe an area of enclosed land, a derivation that can be seen in the names of other British landholdings both in the colonies and in Britain.

The Codrington family genealogy is well documented, and although teasing out personalities and motivations from archival research and publications is interesting, it can also lure one into the trap of conjecture. One thing is certain: like many other planter families, the Codringtons were prone to depart the island as quickly as possible, as Matthew Parker (2011) aptly demonstrates in his book *The Sugar Barons*. We do know from various accounts that Christopher Codrington II, who served as governor general of the Leeward Islands (see chapter 2), was a strong presence on Antigua; one either liked or hated him. Parker describes Codrington II as "colourful, spectacular and controversial," while his son Christopher Codrington III, who went on to establish Codrington College on Barbados (Parker 2011:212), was more easy going.

One of the most remarkable aspects to this story is the fact that Betty's Hope stayed in the family until 1944, making it one of the longest-running plantations owned by one family. Needless to say, Antigua was not their cup of tea, and soon after the tenure of William Codrington in the 1780s, most family members remained in Britain, visiting Antigua occasionally but mostly shielding themselves from the daily grind of running a sugar plantation and the attendant violence and brutality that went with it.

At the heart of this long-term endeavor was sugar. As stimulant foods such as tobacco, sugar, chocolate, coffee, and tea came into the public consciousness, tastes changed, marking a watershed moment in human history from which there was no turning back. Recent attitudes regarding sugar have changed to some degree, and for good reason, as Gary Taubes's scathing critique *The Case Against Sugar* (2016) well demonstrates. Yet in the human craving for sweetness, few are aware of how this commodity was built upon the backs of enslaved Africans. This point was brought home time and again, as tourists visiting Betty's Hope would be surprised as we spoke to them about our work. Many of the things we come to rely on in our everyday lives have some darker aspect to them; the materials used to make the batteries and inner workings of our smart phones, the sweatshop conditions of mass-produced cheap clothing, factory farms, and so on. What archaeology affords us is a means to examine the many processes and complexities involved in the growing, production, and consumption of this alluring commodity. The desire for sugar, regardless of how one thinks of it in terms of its nutritional or sensory effects—as Taubes (2016:30) asks, "Drug or food?"—set in motion a historical trajectory from which we have never fully recovered; The vestiges of colonialism and a slave past are still with us today as we contemplate what the future holds for us and what our place is in it.

Organization of Chapters

This volume is organized by six main research areas: part I (chapters 1–3) provides an introduction to the site; part II (chapters 4–6) offers descriptions of the site's general day-to-day maintenance; part III (chapters 7 and 8) describes agriculture at Betty's Hope; part IV (chapters 9–11) discusses the lives of enslaved Africans and Afro-Antiguans; part V (chapters 12 and 13) locates Betty's Hope in a broader context; and part VI (chapters 14–16) focuses on cultural heritage management of the site.

Chapter 2, by Georgia Fox, provides an overview of the excavation and artifacts recovered from the great house. In chapter 3, by applying geographical information systems to the study of the old maps and using remote sensing, Cory Look reconstructs relationships between key locales, features, and spatial relationships at Betty's Hope, broadening our understanding of changes occurring in the landscape over time and space.

In chapter 4, which centers on plantation maintenance and provisioning, through careful analysis of archival sources, Jennifer Anderson demonstrates the important linkages between Betty's Hope and Antigua to its sister island of Barbuda. In chapter 5, Geneviève Godbout discusses the material culture and provisioning of the great house as it reflects dining and entertaining, one of the chief functions of the plantation house. In chapter 6, Alexis Ohman, who is engaged in a complete analysis of the faunal remains recovered from Betty's Hope, examines mollusk remains from the site as part of the plantation's foodways.

As the key function of any plantation's existence is farming, chapter 7, by Georgia Fox, is devoted to cane agriculture at Betty's Hope. Islands like Antigua provide excellent laboratories for examining environmental change (Kirch 1997). In chapter 8, Suzanna Pratt, Christian Wells, and Anthony Tricarico delve into the long-term impact of cane agriculture at Betty's Hope and surrounding areas, revealing the complexity of environmental transformations.

In chapter 9, James Dator provides context for understanding enslaved life at Betty's Hope by examining the Codrington family's critical role in the Atlantic slave trade. Seeking to unlock the "mental template" of the landscape is often part of the archaeologist's palette in plantation studies (Gosden and Head 1994:113; Hicks 2007:46; Higman 1986). At Betty's Hope, hours of survey work test the notion of the idealized plantation landscape, especially as it relates to the lives of the plantation's enslaved inhabitants. In chapter 10, Georgia Fox focuses on the excavations conducted at the slave village in 2014 and 2015. In comparing the material remains with the archival record, Fox helps situate what life was like for the Afro-Antiguans at Betty's Hope.

Among the most important aspects of enslaved life in plantation settings are the everyday acts of agency and resistance, as evidenced in the hundreds of Afro-Antiguan wares found at the site. Through chemical analyses, in chapter 11, Benjamin Kirby explores the provenience of the Afro-Antiguan wares found at Betty's Hope in comparison to those found on Barbados and Jamaica. By doing so, Kirby assesses the degree of agency that Afro-Antiguan pottery represents among the enslaved population at Betty's Hope.

What lies beyond Betty's Hope? Taking a more macro-scale approach, part V looks beyond the immediate setting of Betty's Hope to include the wider Atlantic world. In chapter 12, Charlotte Goudge examines rum production at Betty's Hope through the excavations of the backside of the still house and the importance of rum in Atlantic trade. To defend their economic interests, the planters relied heavily on military forces for protection. Therefore, no discussion of Betty's Hope would be complete without assessing Antigua's defense system in a sea of hostile relations between European powers as they played out in the Caribbean. In chapter 13, Christopher Waters carefully reconstructs the relationships between Antigua's planters, colonial officials, and the British military in their aim to protect Antigua's sugar assets.

As examples of postcolonial legacies, the last section of the volume, part VI, concentrates on three cultural heritage projects conducted at Betty's Hope. As discussed by Reginald Murphy in chapter 14, the excavations and restoration of the north windmill at Betty's Hope launched archaeological interest in the site and renewed hope in protecting and preserving the island's cultural heritage sites. In chapter 15, Amanda Kramp provides an excellent example of the challenges and joys in creating a museum exhibition at the Betty's Hope Visitors Center. As a tourist destination, Betty's Hope offers multiple opportunities to embark on cultural heritage tourism projects that retain the site's integrity while making the information available to the public. In this vein, in chapter 16, Erin Friedman, Cory Look, and Matthew Brown explore the use of unmanned aerial vehicles in archaeological investigations and heritage management generally and as it relates to the work at Betty's Hope. The afterword concludes the volume with the final days at Betty's Hope.

2

The Great House

GEORGIA L. FOX

As I sat browsing through colorful coffee table books about Caribbean plantation houses, glossy images of ornate porches with scenic views filled the pages, highlighting expansive living spaces adorned with finely crafted furniture. These images are a reminder of the great variety in the homes of former sugar planters, dwellings collectively referred to as "great houses." The Caribbean great house embodied all that was valued and strived for among the British planter elite: status, impeccable taste, domesticity, respectability, and immeasurable wealth. This chapter will focus on our excavations of the great house at Betty's Hope, home of the once powerful and affluent Codrington family.

Nothing above ground is left of the house. All the hardwood floors and antique furnishings are long gone, and not even so much as a teacup has survived. Images of the house's interior are nonexistent, but three images of the home's exterior (1906) surfaced from the Museum of Antigua and Barbuda's archives when we commenced excavations in 2007 (figure 2.1). Locals speculate about the house's demise. Some claim that fire destroyed it, but a search in the local archives turned up nothing to confirm this. Others say that the house's limestone floors and walls were carted away to build a rectory in the village of New Winthorpes, but a visit to the rectory in 2016 yielded no conclusive evidence. Stories aside, archaeological evidence has delivered stone-cold reality; though we have learned a lot in the five years of excavating the great house and kitchen, many questions remain.

Figure 2.1. One of a few surviving images of the Betty's Hope Great House, 1906. Courtesy of the Museum of Antigua and Barbuda.

Betty's Hope: An Outdoor Museum

Betty's Hope plays host to tourists escaping the confines of their cruise-ship cabins and all-inclusive resorts. Visitors can wander the "outdoor museum," consisting of interpretive signage, or they can peruse the visitors center's interpretive exhibits, which focus on the plantation's history and archaeology. Oversight of the property is conducted through the Betty's Hope Trust, which was formed in 1990 by a group of dedicated locals.

This almost 300-year-old plantation, owned by the Codrington family for most of its existence, was once one of the island's most prominent sugar estates. Over the years, the Codrington family amassed significant landholdings throughout Antigua and the Leeward Islands. Their main Antiguan estates included Rooms (1717–1852), Garden Estate (1755–1820?), Cotton New Work (pre-1738), Tuits (1760), Cotton (ca. 1764), The Folly (?–1773), Bolans (1772–1815), Jennings (1772–1815), and Clare Hall (Skerretts 1814–?). Additionally, the Codringtons leased Barbuda and possessed the small, steep Guiana Island (1812–1929), off the northeast coast of Antigua (Devonian Foundation 1988).

The Early Years

Not much is known about the early years of Betty's Hope and its first owner, Christopher Keynell. Arriving in 1652, Keynell was sent from England to Antigua to deal with the fledgling colony of about 1,000 immigrants (Dyde 2000:15, 17; Oliver 1894–1899:II:123). Things had not gone well, as struggling colonists eked out a living growing tobacco, cotton, ginger, and indigo (Dyde 2000:17). When Keynell acquired Betty's Hope is unknown, and even less is known about the acreage he acquired. Oliver (1894–1899:II:124) states that "30 mens land by sale from Mr. Henry Austin [? Ashton] 17 June 1664, also 42 acres by pattent from Governor Austin to Clement Allen and George Rawlins 27 Feb. 1656, and confirmed to Colonel Christopher Keynell by sale 11 May 1657, also 74 mens land bought by Colonel Christopher Keynell."

What is certain is that Keynell served as governor from 1652 to 1660 and then died in 1663. Following Keynell's death, Betty's Hope passed to his widow, Joan, who subsequently married John Hall (Nicholson 2001:45; University of Florida 1994:82), who lost the estate after fleeing with other estate owners to the nearby island of Nevis during the French invasion of 1666–1667. Hall's misfortune derived from parliamentary land reforms devised to encourage and "promote the settling of land" by planters (NACO Acts of Antigua; Sheridan 1974:190). Revoking these landholdings paved the way for planters like Christopher Codrington II, who was granted all 725 acres of Betty's Hope in 1668. An item in the National Archives records the "Plantation formerly the Widow Joan Kennell's [?] Betty's Hope, being granted unto Collonell Christopher Codrington" (NACO Acts of Antigua). Oliver (1894–1899:II:125) offers an account of what happened, along with some tantalizing clues about the early history of the estate and the attitudes toward women landowners at this time:

> Dame Joan Hall, Widow and Relict and Ex'trix of Colonel Christopher Keynell, late of Antigua, stating that she possessed a plantation called Bettye's Hope for 14 years, *circa* Anno 1667, and the French invading the island, she went for safety to Nevis, leaving 60 negros behind, all of whom were taken or killed, and the estate ruined. Later the neighbours made a garrison of her house and burnt down her sugar works, so that their security became her ruin. She returned in

> 1668 and repaired the buildings, but it so happened that William, Lord Willoughby of Parham, then Captain-General, brought with him one Colonel Codrington, to whom he gave her plantation, alleging that it was too great a quantity of land for her. Not yielding Colonel Codrington the anticipated profits it's offered for sale.

The Codrington Legacy

Christopher Codrington II, a planter's son from Barbados, did not arrive in Antigua until 1674 and did not permanently settle at Betty's Hope until 1689, when he assumed the role of governor general for the Leeward Islands (Sheridan 1974:193). Once Codrington II assumed active involvement in Betty's Hope, a new era of the Codrington dynasty began, lasting until the plantation was sold in 1944. Although the Codrington family became the leading planter family of Antigua, there is a common and long-held misconception that Codrington II began the first sugar plantation on Antigua—an implausible claim considering that there were other well-established planter families, such as the Oliver, Byam, and Martin families (Lowe 1951:1, 5; see Oliver 1894:I). What Codrington II did contribute was the requisite knowledge and expertise in cane agriculture that he had gained during his years in Barbados. Historian John Oldmixon (1708:181) claimed that Antigua "has ow'd most of its flourishing Condition since to the Care and Intent of Col. Christopher Codrington . . . having great Knowledge and Experience in the Sugar Plantations. . . . Others following his Example, Antego [Antigua] thrive."

Christopher Codrington II's origins trace back to Gloucester. His father, Christopher I, left England and arrived in Barbados sometime in the 1640s or possibly earlier (Davis 1887:144; Gaspar 1985:66; Harlow 1990:6; Lowe 1951:1). Barbados, where Codrington II was raised and where he learned the sugar business, was the most promising of the West Indian colonies and the home of English sugar production. Unlike James Drax and some of the early Barbadian planters, Codrington I had aristocratic cachet; his pedigree could be traced back to Henry V at Agincourt (Harlow 1990:1; Parker 2011:38). Regardless, these early Barbadians were pioneers in their own right, facing formidable challenges in the early days of settlement (Harlow 1990:6). Their circle was intimate and insular, with intermarriage common, as was the case of Codrington I, whose marriage to Frances Drax, sister of James, merged the two powerful families, their union producing Christopher Codrington II, born in 1640 (Harlow 1990:9).

After leaving Barbados, Codrington II led a colorful and productive life, leading military campaigns and amassing a huge fortune. When Codrington II died in 1698, Betty's Hope passed to his son Christopher Codrington III. Educated in England, Codrington III was "far more cultured and less of a libertine than his father" (Dyde 2000:40). As the next governor general, Codrington III carried out his duties until 1704, leaving Antigua permanently for Barbados in 1707. His departure signaled the beginning of absentee ownership at Betty's Hope. After Codrington's death in 1710, Betty's Hope passed to William Codrington I, 1st Baronet of Dodington and Christopher Codrington III's cousin. In Codrington III's last will and testament, executed at Betty's Hope, he left "to my Cousin Leiut. Coll Wm Codrington my plantation at Betty's Island in Antigua" (Harlow 1990:218).

Beginning with William Codrington I, daily operations at Betty's Hope would be supervised by an ever-changing parade of managers and attorneys, with Codrington family oversight being administered from the coddled life of Dodington Park in Gloucestershire. Following the death of William Codrington I in 1738, the estate passed to William II, 2nd Baronet, who owned the estate until his death in 1792 (Codrington 1779–1782). It was during William II's oversight that Betty's Hope underwent important modifications and improvements in response to the growing sugar market and increased efficiency (see chapters 3, 7, and 12). Unfortunately for William III, who was disinherited by his father in 1789, Betty's Hope and William II's other estates went to his nephew, Christopher Bethell Codrington (b. 1764).

Inheriting considerable wealth, Christopher Bethell followed through on his uncle's dictate to "reside for six months after my decease on my estate in the West Indies to learn planting." Following Bethell's death in 1843 (Thorne 1986:201; Wilson and Mackley 2000:325), Betty's Hope was inherited by Christopher William Codrington (1805–1864) (Nichols 1867:7), followed by Gerald William Henry Codrington (1850–1929), with final ownership by Christopher William Gerald Henry Codrington (1894–1979), who sold the estate in 1944 to Antigua Syndicate Estates Ltd. (Baronetage of England, Ireland, Nova Scotia, Great Britain and the United Kingdom 2016; Museum of Antigua and Barbuda 2005).

The Great House and Kitchen Excavations, 2007–2012

Archaeological excavations at Betty's Hope, including the great house, have been reliant on two maps in the Codrington archive: the maps of 1710 (NAAB Codrington Papers:P3) and 1755 (NAAB Codrington Papers:P10), which feature the buildings and surrounding lands at Betty's Hope and adjoining Codrington estates (the old maps will be discussed and referred to in several chapters throughout the volume). Both maps show the great house centrally located with a separate kitchen and nearby "complex" of support buildings as shown in figures 2.2 and 2.3. To the west was the sugar factory, comprising the still, curing, and boiling houses and the double windmills. To the north and west was the manager's house, and further into the interior were the animal pens, water catchment areas, and ponds, as indicated by the site plan (figure 2.4). Comparing the maps to the archaeology has allowed for determining important changes and transitions in the plantation's history.

Consulting with Reginald Murphy, the island archaeologist, we decided to launch excavations at the great house because of its central location in relation to the rest of the 50-acre site. When excavations of the great house began in 2007, the looming question was where to start. The

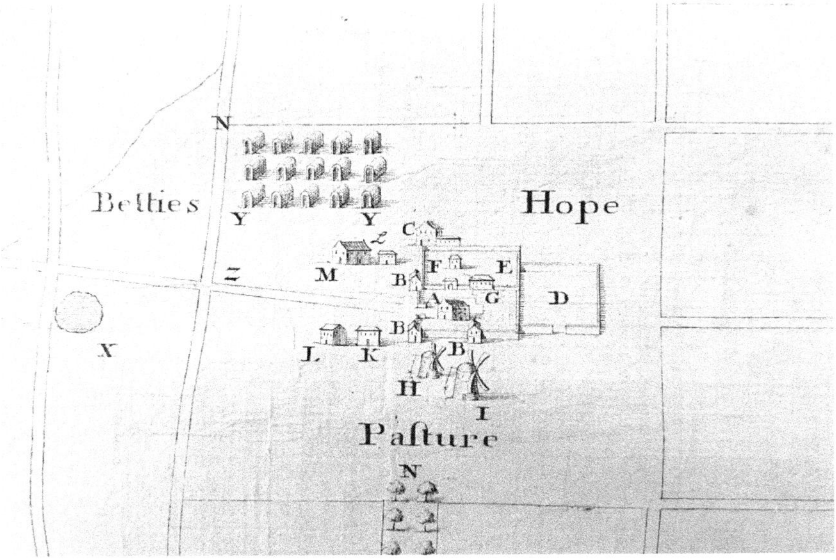

Figure 2.2. Close-up of the Betty's Hope complex, 1710.

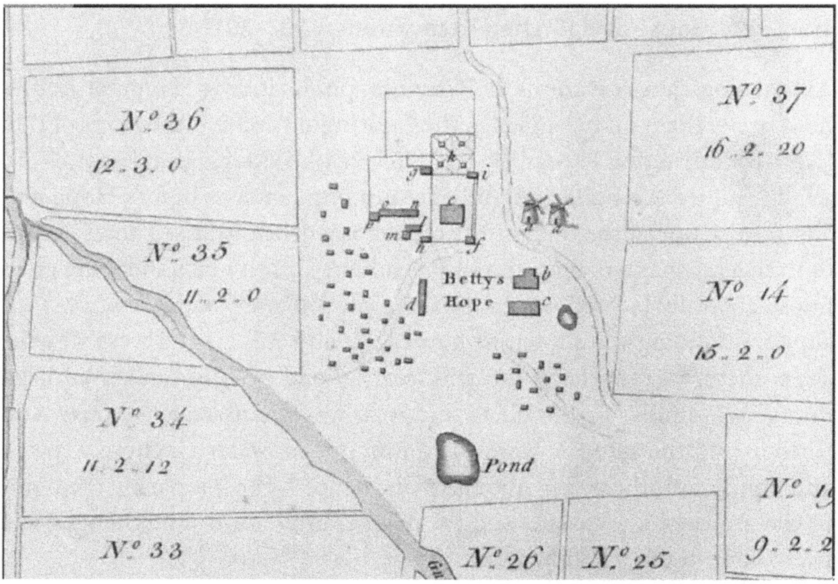

Figure 2.3. Close-up of the Betty's Hope complex, 1755.

1755 map depicts Betty's Hope as enclosed by a parallelogram-shaped wall, with a small building located at each corner of the wall. These support buildings included an overseer's office (northwest), a bookkeeper's office (northeast), a doctor's office (southwest), and a tradesman's office (southeast corner). Outside the wall and to the north was a complex of separate service buildings, including a servants' quarters, a sickroom or "hot house for the sick Negroes," a blacksmith shop, and a stable (now the visitors center).

Consulting the maps as well as the old exterior photos, we began excavations at the southwest entrance to the great house, across from the windmills. This strategy allowed us to conduct excavations systematically in a northerly direction for the next five years, following the discovery of the first features at the end of the 2007 summer field season. Areas were laid out in 2 m^2 excavation units, using natural stratigraphy, as indicated by changes in soil color and texture (in the volume, archaeological measurements are given in metric units, but discussions related to archival sources are expressed in imperial and US feet for comparative consistency). It should be noted that although the great house soil horizons have been affected by some site disturbance and fill, the integrity of the site was

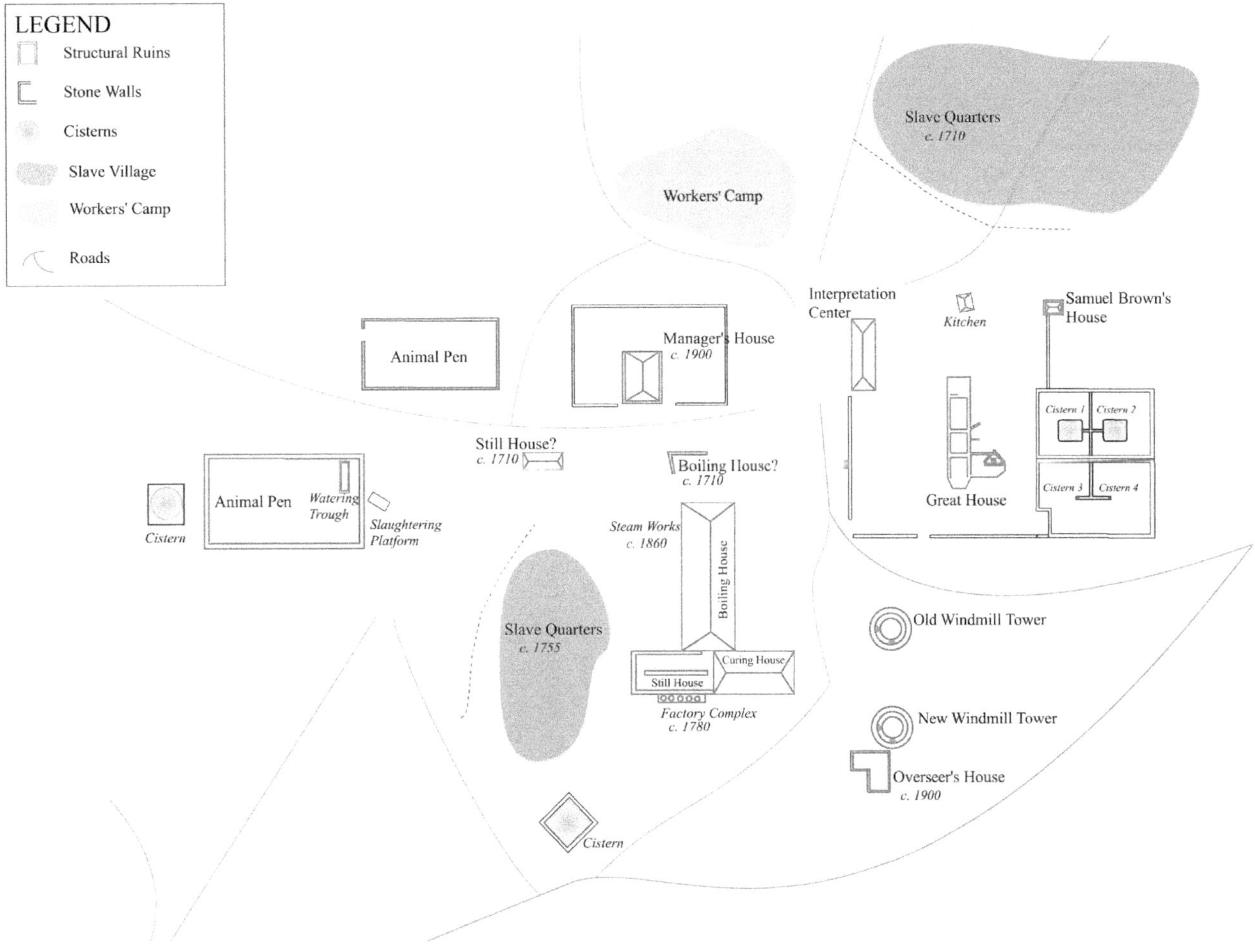

Figure 2.4. Site map of Betty's Hope Plantation. Map by Cory Look.

Figure 2.5. Exposed brick floor and foundation walls uncovered during the Great House excavations. The area shown features the brick walkway between the Great House and kitchen.

not compromised. All excavated soil was screened using 3 mm wire mesh to retrieve small artifacts and ecofacts.

The great house excavations comprised two overlapping phases: the house location proper, including the walkway between the kitchen and house (2007–2010), and the kitchen area (2010–2012). A separate kitchen made sense in the hot climate and to protect against the ever-present danger of cooking fires and unwanted cooking smells. The kitchen was two stories and probably accessed through a brick archway that existed between the house and the kitchen (see figure 2.5). Large plantation kitchens like the one at Betty's Hope were indicative of great wealth and status, particularly if they included a large central hearth and stone walls (Hobson 2007:282). Although the central hearth is long gone, the foundation for a smaller chimney was discovered during the kitchen excavations.

To locate the kitchen, a pedestrian survey was conducted in 2009 utilizing a transect at 15 m intervals going from east to west, as well as a series of 12 shovel test units (STUs) of 1 × .5 cm. The STUs were excavated to 20 cm levels, resulting in the discovery of a red-brick floor as well as part of a cut limestone floor, which were interpreted as being the archway and kitchen areas, respectively. Previously, in 2008, an unusual feature

Figure 2.6. Mystery feature recovered in 2008, possibly the foundation of a small smoke house. Courtesy of the National Archives of Antigua and Barbuda.

was discovered and excavated, which also helped locate the kitchen—a circle of brick with plaster-lined walls at a shallow depth of 30.48 cm and a diameter of 82.5 cm (as shown in figure 2.6), in the middle of which is a square-shaped brick hole that looks like it once held a post.

One explanation for this mystery feature is that it served as a "meat hanger" located at the back door of the kitchen, in which case it may have been filled with water to ward off pests. A similar feature is located inside a storage closet in the Dockyard Museum's gift shop. Another possibility is that this feature is the foundation of a small smokehouse for curing meat. A close look at the 1906 image of the backside of the kitchen shows a beehive-shaped structure, which might have been the smokehouse. The smokehouse at Boone Hall Plantation in South Carolina bears some similarity, also made of brick, with slate roof tiles. If this is the case, the circle may have been filled with wood for smoking meat. There is a precedent for this, as British colonists introduced and constructed smokehouses in seventeenth-century Virginia, particularly in the smoking of ham. Over time, the smokehouses probably integrated a combination of British and West African traditions, some of which may have originated in the West Indies (Edwards 2011).

Garden and Great House Support Buildings

Garden and Walkway

The English penchant for a prominent house with a walled garden typifies the arrangement at Betty's Hope. Little is known about the enclosed garden at the great house. Of the wall itself, only a small section exists on the south side of the house, but its former appearance can be surmised by the extant walls surrounding the four large cisterns just outside the great house to the east. This type of wall seems to be typical of eighteenth-century wall design and construction on the island, some of which can still be seen today. The walls are composed of loose stone, plastered on the outside and capped with a round-like sloping top. Architect Daphne Hobson (2007:290) attributes a Baroque influence at Betty's Hope, exemplified by formal gardens and walkways. A former formal walkway lined by tamarind trees extended outside the walls of the great house to create a sense of grandeur to the eastern side of the property. Labeled as the "Tamarin Walk" on the 1710 map (see chapter 3), most of this walkway is now obliterated on the landscape, but a current path still follows the original layout. Utilizing the 1710 map with satellite imagery and hand-held GPS, we conducted a pedestrian survey in 2016, discovering that the path directly connects Betty's Hope to the adjacent Garden Estate. Even more exciting was the discovery that this walkway led to the provision grounds located on the 1710 map, now the present-day area of the Potworks Dam reservoir.

Support Buildings

In regard to locating the support buildings north of the great house, extensive shovel tests were conducted in 2011 and 2012 by Geneviève Godbout and Katherine Davis and in 2013 by Godbout (see chapter 5). On the 1710 map, these support buildings are referred to as flankers. Surveys of the area north of the great house area were hampered by thick vegetation until a backhoe operation carried out by the Antiguan government in the early spring of 2014 exposed much of it. The results were startling: the service buildings were much closer to the great house than previously thought, as the heavy vegetation was deceptive. The recovery of bridle bits and iron slag, as well as horseshoe nails in 2014, also suggests that we were much closer to the former blacksmith shop and stable depicted on the old maps.

As maps are only a snapshot in time, we cannot expect a perfect match between eighteenth-century idealized maps and on-the-ground reality. Although our archaeological excavations and surveys do account for some foundations and features consistent with the old maps, there are other features that do not appear on them. Additionally, none of the support building foundations can be positively identified as to specific function.

In 2011, an unexpected but hoped-for area was discovered: a deep privy that was located adjacent to the great house near the south-side entrance. The privy was excavated to over two meters in depth, but because collapse posed a danger, excavation was terminated. Stratigraphy was clearly delineated and dated by the presence of ceramics and clay tobacco-pipe fragments. The artifacts are in line with what one would expect from a site occupied from the late 1600s to early 1900s. To date, it is the only privy found on site.

The Material Culture of the Great House

Architectural

There are few remnants to suggest that the great house was two stories, but the old photos substantiate this, as does the discovery of stone post insets in the great house floors, which would have supported post and beam construction for a second story. Stone impressions for door posts were also uncovered. The first floor probably served for messy domestic chores such as laundry. Some great houses also had baths and bathing areas in the cellar (Agnes Meeker, personal communication, June 14, 2007).

The main living area, on the other hand, was relegated to the second story to take advantage of the cool sea breezes. The once large roof was covered in Welsh slate tiles, which have been found on site. The foundations, walls, and floors of the great house, kitchen, and outbuildings were constructed of hand-carved blocks of limestone that were held in place via limestone mortar. The walls were traditionally faced with plaster, and in some instances, the plaster was painted different colors, as indicated from the archaeological finds.

The material culture of the great house architecture comprised voluminous quantities of hardware, including wrought and machine-made iron nails, window glass, lead caulking, hinges, plaster, chunks of mortar, brick (crude and uniform), red and terracotta tiles, Welsh slate roof tiles, a few door locks, and iron shutter pins; Douglas Armstrong (1990:202–205, 270) reports finding similar architectural material at the Drax Hall excavations in St. Ann's Bay, Jamaica.

Miscellaneous items relating to architecture also include a section of a glazed ceramic water pipe found extending outside the house, possibly used for water drainage. Two extant iron standpipes formerly located on the bottom porch were useful guides, as they appear in the early photographs. Finally, several attempts were made to determine the composition and age of the window glass through scientific analyses, but these efforts were not fruitful.

Artifacts

In excavating the great house and kitchen, we expected the material culture to reflect a domestic sphere. Glass bottle rims, bases, and fragments of rum and wine bottles were recovered, but there was little evidence of stemware. Only a few medicinal bottles were found at the great house, and cutlery was also noticeably absent. Identifiable fragments of clay tobacco pipes were recovered but were not in great abundance.

Categorization of the Betty's Hope artifacts was based on standard historical archaeological nomenclature and categories and subcategories for artifacts as well as for provenience, date, and measurement information. Dating the great house artifacts was based on standard relative dating, placing the majority of artifacts squarely in the mid- to late eighteenth century range, with outliers on either side. Artifacts were statistically calculated using IBM SPSS Statistics and converted to graphs for each artifact category. The distribution of common imported identified ceramics found during the great house excavations includes (in number of finds) pearlware (159), Chinese porcelain (94), whiteware (74), salt-glazed stoneware (69), transfer printware (54), Delftware (48), creamware (42), Staffordshire slipware (11), yellow ware (3), mocha ware (3), luster ware (2), annular ware (1), and agate ware (1).

Utilizing the project's completed database, a breakdown of general artifact categories for the great house includes clay tobacco pipes, bottle glass, imported ceramics (predominantly British), personal items (mostly buttons), firearms (musket balls and flint), and architectural elements such as nails, window glass, stone, plaster, and hardware. In 2009, a significant cache of children's toys and sewing notions was found, suggesting that they might have dropped through the upper porch living area. The sewing-related artifacts include hooks and eyes, needles, buttons, thimbles, and pins. The toys found reflect the Victorian era through a collection of marbles, parts of a tiny chess set, and a miniature man with a top hat made of painted metal.

Other artifacts from the great house reflect domestic life, especially cooking and laundry. During the 2009 summer field season, parts of a cast-iron stove were recovered (three feet, one burner, and a handle). A few feet away, the stove's label was found, identifying the maker as the Albion Lamp Company of Great Britain, 1896.

As laundry was probably done in the cellar of the house, in 2010, the top section of a cast-iron hand-cranked washing machine mangle was recovered, with the manufacturer's name bearing "P. Johnston 290, Oxford St. London," reflecting one of the great inventions of the Industrial Revolution. Other laundry-related materials include bluing balls to make clothes look whiter and part of a clothesline and pulley-gear system for hanging laundry.

Small parts of furniture hardware were also discovered, such as tiny drawer or chest handles, as well as a substantial ornate iron door knocker. In terms of the interior of the house, the Codrington Papers provide some specifics of the house's former opulence. As early as 1715, an inventory of purchases includes pewter plates and candlestick holders, a large silver punch bowl, forks, spoons, and knives, one diamond ring, French and Spanish pistols, one gold chain, wine, and sheets and pillowcases (NAAB Codrington Papers:C2).

Acquisitions for the house in 1781 by estate manager Richard Oliver include cedar chairs, a sideboard, a pair of tables, an eight-day clock, and a mahogany tea table, bookcase, and bureau, with mahogany being the preferred choice for high-end fashionable furniture (Anderson 2012:11; NAAB Codrington Papers:C12). Other artifacts included chunks of sugar pot and coarse earthenware, but almost no Afro-Antiguan wares, the locally made pottery. Very few items of personal adornment were found, with the exception of buttons and small beads. In 2008, seven coins were recovered, including two copper coins dated to 1886, picturing Queen Victoria; four three-pence silver coins dated to 1904 with the visage of Edward VII; and a silver coin dated to 1897 featuring the head of Queen Victoria.

Interpretation of the Great House

The local lore of the great house meeting a fiery end is intriguing but unproven. What is known is that by October 1936, the house was referred to as the "old mansion" and described by Codrington attorney W. E. Walwyn as being "very difficult to realize that the house was in good order and inhabited only nine years ago, and is now in ruins" (NAAB Codrington Papers:C67). Less than a month later, Walwyn reports that he realizes "how sore the owner must feel over the destruction of the Big House, and I trust that I'll have the good fortune to restore this." This implies that the house was no longer standing by 1937, and good intentions aside, restoration plans were never realized.

Compared to other plantation houses, Betty's Hope appears to have been a modest affair. Perhaps this can be explained by its reliance on absentee landlords since 1707 and by much of the wealth having been diverted to the grand Codrington estate at Dodington. As planters established themselves in the early English settlement of Barbados and Antigua, housing was more basic and rudimentary, but within a 20-year span, large homes began appearing on Barbados. Therefore, it would not be unreasonable to assume that the first dwelling at Betty's Hope was a much simpler affair than the dwellings built during the Codrington years and that it burned down when Joan Hall fled to Nevis. It was only after 1650, when sugar production began to increase, that plantation houses on Antigua were built of stone, wood, and brick (Hobson 2007:184).

Because Christopher Codrington II did not permanently settle at Betty's Hope until the 1670s, the great house was possibly more modest until he became governor general. It appears that he was planning to build the plantation's structures out of stone, more befitting a man of his stature, and that his plans included a grander house. Unfortunately, the project was temporarily interrupted by an earthquake. In a letter addressed to the Lords of Trade and Plantations dated June 4, 1690, he states:

> Soon after despatch of my last there was a terrible earthquake, which laid some of our buildings in rubbish and killed some persons. Scarce any stone-work in these Islands has escaped without damage, and I myself am a loser to the value of £2,000. The great earthquake was on Sunday, 5 April, between four and five o'clock in the afternoon; for a month afterwards we had almost daily shakes, and even now there passes not a week without some tremblings. (Fortescue 1901 [1690])

After Christopher Codrington II died in 1698, all responsibility passed to his son, Christopher Codrington III, who did not share his father's enthusiasm for running a plantation. When Codrington III, haunted by a difficult military service and serious illness, left Betty's Hope in 1707, he had little inclination for maintaining a fine home at the plantation. Although the house had become dilapidated by 1710, the fine craftsmanship of the carved limestone blocks and thick walls suggests that it had been built with substantial means that only a Codrington could afford. That the great house appears on the 1710 map suggests that the original limestone walls and foundations at least date to this period.

Interestingly, there is scant mention of the great house in the Codrington Papers, and what little is there relates to repairs on the house, which appear to correspond to overall plantation expansion and improvements in the 1710s, 1750s, and 1780s and to repairs stemming from natural disasters. For example, during the period of 1741–1751, in a series of letters from attorney Benjamin King to Sir William Codrington II and Lady Codrington, King reports that "the dwelling house tho lately repaired by Coll. Gunthrop at great expense. The roof is so flat a pitch it leaks all over, so it must soon decay. I have seen it in a season of rain, when there has not been a dry board in all the floor, all the out houses, and cestrines [cisterns] very much decayed and in wretched order" (NAAB Codrington Papers:C5).

In another letter dated April 22, 1782, William Codrington II laments to manager Richard Oliver the ongoing "evident state of expense in Buildings at Betty's Hope" (NAAB Codrington Papers:C12). In an 1844 letter from James Salmon to Christopher Bethell Codrington, Salmon states that "the Dwelling House at Betty's Hope is in thorough repair" (NAAB Codrington Papers:E23). In 2011, architect Scott Mulrooney provided us with an AutoCAD drawing of the areas excavated from 2007 to 2010, which was helpful in visualizing the general outline of the great house, but we were unable to assess the dwelling's original size. Ultimately, about 95% of the house and kitchen foundations were exposed. As with many aspects of archaeological inquiry, we will never know the full extent of the house, its furnishings, its smells, sounds, and even its inhabitants; we can only imagine what it was like. The truth remains elusive, casting a long shadow over the mysteries of the plantation's past.

3

Reconstructing a History of Plantation Spaces at Betty's Hope

CORY LOOK

Arguments have been made that the decline in Antigua's sugar production was a byproduct of poor land management practices, depletion of soils, and outdated agricultural techniques and technology (Abbott 1964:1; Campbell et al. 1992; Ragatz 1928; Ward 1978). Recent work at the site of Betty's Hope Plantation suggests that not all plantations were susceptible to poor decision making (Pratt 2015; Wells et al. 2018b) and that some plantations may have superseded social-ecological constraints despite the sugar industry's inevitable decline. This chapter attempts to contribute to this discussion by focusing on key aspects of land-use and landscape change at Betty's Hope, particularly during the eighteenth century, when the plantation experienced rapid growth and economic prosperity. It is the intent of this research to understand the impact on the enslaved class through examination of changes to slave quarters' spaces and provisioning grounds by managers of local resources and land use.

Recent innovations in remote sensing techniques, aerial imagery, and spatial software have allowed new bodies of data to be generated on slave landscapes that were not previously available. To achieve this goal, the identification and mapping of walking paths, roads, and boundaries were used to help quantify changes to lands that were not part of the estate's original accounting ledger. These new bodies of data were used to synthesize a history of recorded and unrecorded spaces at Betty's Hope, with the intention to shed light on whether land management decisions and technology played a role in the plantation's success.

From the late sixteenth century, estate maps in England were tantamount to individual landowners, each making up a single rural economic unit. Accuracy and clear territorial boundaries were essential to a market-driven agricultural system (Buisseret 1996), as they resulted in spaces that could be accurately recorded and revisited (Pickles 2012). This type of inscription may have kept land in a constant state of development, and that land potential was measured by potential yield. However, as planters were a class of power elites, the lines they chose to draw, erase, and redefine had direct impacts on those enslaved and those laboring on sugar plantations. Therefore, the integration of archival maps is not just about understanding what they show or how things are represented, but rather in what the map accomplishes (Wood 1993:56).

Remote Sensing, GIS, and Cartographic Methodologies

Beginning in 2011, a broader landscape component was added to the ongoing excavations at the great house, support structures, and slave quarters at Betty's Hope. The initial landscape focus was on identifying the variety of lived spaces that planters, laborers, and the enslaved shaped throughout the plantation's history.

This involved the creation of a "fuzzy" base map that could integrate loosely oriented estate maps with well-defined chronologies and geospatially accurate features identified from archaeological investigations with chronologies based on the known history of the plantation, relative dating through artifacts recovered, and stratigraphy. The ability to scale up or down was necessary to test temporal relations with their spatial counterparts, from excavated units to offsite landscapes. The use of geographic information systems (GIS) within archaeological investigations has become standard practice and is ideal for handling multiscaler datasets; however, these databases are often heavily biased toward the researcher's questions and objectives, which are not often made explicit when employing GIS software.

The science of cartography was an attempt to model reality while at the same time communicating spatially driven information to a specific audience. GIS software integrates cartography while adding a complex suite of tools geared toward the study of space and, to some degree, time. The capabilities of GIS software, such as GRASS, QGIS, and Esri's suite of GIS software, go beyond the limitations of cartography and integrate spatial statistics and modeling. Both GIS and traditional cartography are ultimately plagued by the same condition: maps are often not a reflection of the experienced realities of people living inside map boundaries, but rather are more idealized landscapes. This is especially true of Betty's Hope, a case of the colonial mindset.

This can be compounded by a growing recognition from both cartographers and GIS practitioners who have recognized the subjective role they play within socially constructing realities, and more attention, particularly from critical GIS, has demonstrated the pitfalls and dangers of not making one's intentions clear from the start (Pickles 2012). With this in mind, the Betty's Hope Archaeological Project has placed an emphasis on analyzing two estate maps, one commissioned in 1710 by William Codrington I and drafted by James Porter (NAAB Codrington Papers:P3) and another from 1755 drawn by Samuel Clapham and commissioned by William Codrington II (NAAB Codrington Papers:P10).

The initial idea was to test the accuracy of these maps, to study their intentions, and to engage the discursive framing of these documents alongside findings from archaeological investigations. These findings were then used to explore the contradictions and tensions that formed during the estate's period of economic success and form a working understanding of post-1755 land-use and landscape changes that occurred during its eventual collapse.

Situating the 1710 and 1755 maps required an extensive amount of pedestrian surveys and shovel tests, combined with new remote sensing techniques. The use of drones for low altitude aerial photography has become increasingly popular within archaeological survey work (as will be discussed in greater detail in chapter 16). A set of aerial images broken up across the island of Antigua was made available thanks to Michelle Henry of the Museum of Antigua and Barbuda. Archived aerials existed for each decade since the 1940s, and when combined with low flying aerial imagery (drones), this methodology proved to be an effective combination.

While the archival imagery showed changes in ground cover and vegetation, the images were often too blurry or of too low a resolution for making sense of ambiguous geometric patterns or archaeological features. The use of drones allowed our team to target particular areas of interest prior to archaeological survey work. For these sensitive areas, sets of images were taken during a variety of conditions throughout the day, including overcast, after rainfall, and during periods of intense shadowing.

Periods of extended drought, such as the summer season of 2014, were particularly valuable for identifying boundaries and paths, while rainy seasons with thick vegetation were useful in identifying enclosures, water catchments, structures, and agricultural land-use areas. Roofless structures—partially intact foundations—and wall enclosures often resulted in geometric configurations that were distinguishable from vegetation shaped by local grazers. When areas of dense vegetation inhibited the collection and identification of ground control points, we constructed targets that were visible from low flying aerials and relied on total station points for georeferencing.

Walking Paths, Roads, and Boundaries

Walking paths, roads, and land-use boundaries formed some of the more stable and persistent landscape features within the case study area, proving invaluable for establishing initial ground control points needed to situate and begin testing archival information. This same accuracy meant that georeferenced maps could also be used to measure changes in land use and calculate how these changes may have affected those living within the map's boundaries.

Observing the 1710 estate map, many of the land parcel boundaries, labeled as "pieces," align quite well against aerial imagery (figure 3.1, 1710 Overlay). These land parcel boundaries doubled as roads, and their linear and rectangular forms helped accommodate carts and wagons hauling hogsheads to their point of distribution (Higman et al. 1998; Higman 2001). These same geometric forms were critical in identifying non-structural features such as old walking paths and boundaries for slave quarters. The earliest documented location for the slave quarters is associated with the 1710 map, placing it northwest of the great house; however, the same map with its accurate portrayal of agricultural land is poorly scaled when depicting plantation structures. As will be discussed in the section on buildings, while structures are far easier to identify and map, they appear to be more susceptible to modification, relocation, and abandonment, making it more difficult to connect with subsequent estate maps.

In the case of the slave quarters, there were no standing structural features to help approximate the early slave quarters area. As can be seen in figure 3.2, the slave village (Y) abuts a cart and carriage road to the west, while a walking path to the north runs east to west. Using the 2009 aerial

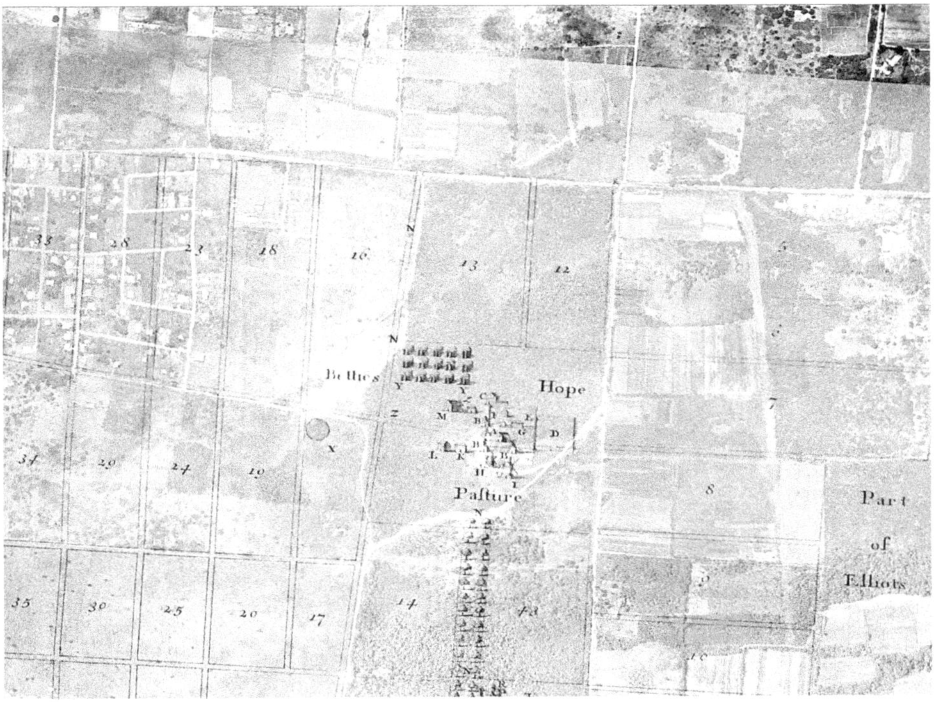

Figure 3.1. The Codrington 1710 estate map georeferenced and overlaid atop an ortho-rectified aerial image obtained using UAV (drone) survey and structure from motion modeling.

image acquired during drought conditions, the walking path can be seen in figure 3.3 starting just east of (N). This same road is easily walkable today, and it is aligned with a series of large tamarind trees that seem to demarcate portions of this path. Aside from the two tamarind trees within the main sugar complex, the nine other standing trees occur along the boundaries of the 1710, 1755, and post-1755 enslaved quarters.

These walking paths are to some degree maintained by local grazers such as goats and sheep that continue to make use of these historic routes. However, more of the land surrounding Betty's Hope has been enclosed by large metal fences that are beginning to restrict access. It is uncertain what impact this will have on old walking paths, although recent visits to the site suggest that they may be susceptible to being lost over time. Neither the 1710 nor 1755 map indicates that the industrial sugar complex encroached or extended into this area; therefore, it is reasonable to assume

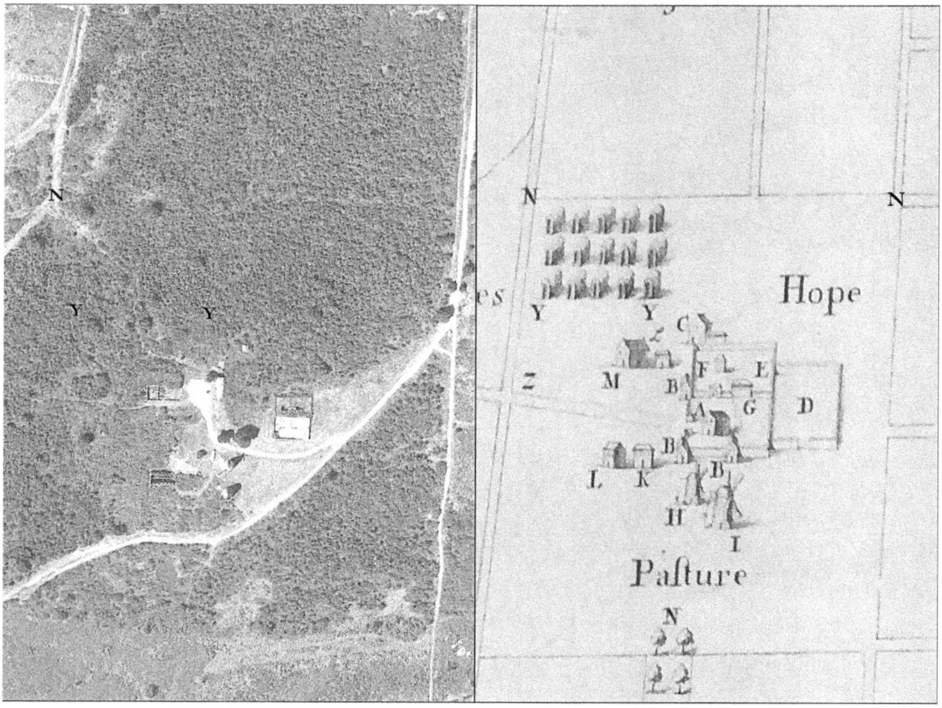

Figure 3.2. Low altitude aerial acquisition used to analyze and test sugar works layout commissioned in 1710 by the Codrington family. Note the visible land boundaries, walking paths, and roads still visible.

that this northern walking path would have marked the limits of the slave quarters and would have been used frequently by the enslaved peoples. What was not identifiable were the hidden walking paths that slaves may have had access to. To remain hidden, these paths would have had to change frequently to avoid leaving semipermanent changes in ground cover.

Some walking paths were more difficult to identify, such as the Tamarin (tamarind) Walk, as indicated in the 1710 map. This walkway originated along the northeastern edge of the 1710 slave quarters, and without any continuity, it reappears just south of the windmills at Betty's Hope, moving south toward the garden (figure 3.3, Tamarin Walk). It is possible that the walking path just northeast of the slave quarters continued along the western edge of the plantation or perhaps ran directly through the heart of it. The walking path just south of the windmills was largely

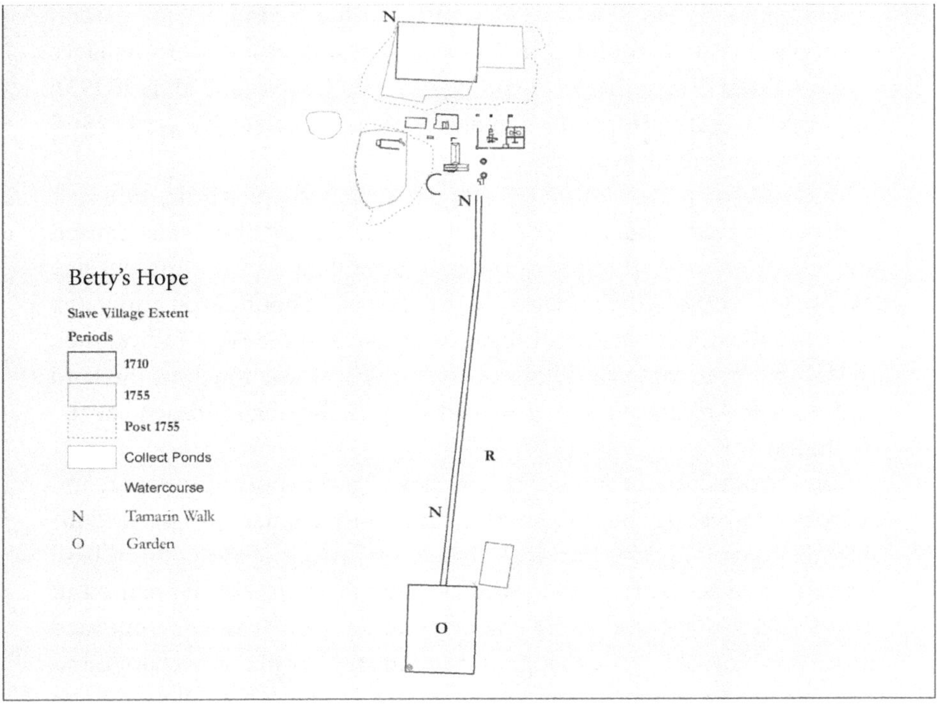

Figure 3.3. Map of the Tamarin Walk based on the 1710 estate map. This includes the Garden, which by 1755 becomes part of the Garden Estate boundaries. The watercourse and collect pond are still present today, with some changes to the patterning possibly due to recent land modifications.

overgrown, as it ran through parcels of fenced-in agricultural land and was additionally left fallow for some time. The Tamarin Walk was no longer labeled or demarcated in the 1755 map, and the garden was replaced by the Codrington-owned Garden Estate. Archival imagery, however, shows that this walking path was still utilized and maintained at least until 1968 and that it led to a rectangular enclosure consistent with the original 1710 garden outline. The southern extent of the Tamarin Walk is still intact and is used by local farmers to move agricultural equipment around. It should be noted that there were no tamarind trees aligning this walk; however, a local variety of tamarind which does not fruit was growing in the areas left fallow. These local tamarind trees are considered to be more of a weed and are found all around the plantation. The southern remnants of the Tamarin Walk led to the rectangular enclosure consistent with the 1710 garden boundary. As described by Charles Rose Ellis in 1797, the garden was a place that surrounded the houses of plantation workers' villages and was distant from the provisioning grounds (Higman et al. 1998:193); although for Betty's Hope, the Tamarin Walk led south to the garden and into the provisioning grounds.

The land-use pattern of Betty's Hope has changed very little, although evidence suggests that recent abandonment during the 1960s caused severe erosion around much of the agricultural land in the present day (Wells et al. 2018b). The remarkable persistence of boundaries and paths is consistent with other British sugar plantations in the West Indies, such as Montpelier in Jamaica (Higman et al. 1998), where much of the land continues to be farmed and managed despite the abandonment of the plantation.

The introduction of modern plows has begun exposing the marl and bedrock, suggesting that modern farming practices may not be sustainable for much longer in this area; increased fencing of agricultural land has restricted access to grazers who help to maintain walking paths and abandoned roads. These recent changes make a strong case for continued documentation of the site as it goes through another phase of change and to understand past land-use and management practices that allowed these lands to be sustainably farmed for almost three centuries.

Landscapes of the Enslaved

Archaeological research has begun highlighting the role that enslaved Africans in the British West Indies played in planning and shaping the spaces around them (Armstrong 1999; Hauser 2011; Mullins 1995). Their creation of alternative economies and direct participation in capital exchange at the plantation level is best described in the *Select Committee on the Extinction of Slavery Throughout the British Dominions*, originally printed in 1832. Slave quarters or villages in the British West Indies, particularly as sugar production intensified in the late eighteenth century, were often inconsistent with the planter's vision of the imagined colony, and as Douglas Armstrong states (1999:179), "While the planter controlled decisions concerning which area was occupied, there is no indication within it of the planned community organized in the previous century." The following section seeks to better understand the landscapes of African Antiguans through the spatial analysis of the quarters and the provisioning grounds, both areas of resistance and areas under the direct control of the enslaved (Mullins 1995:300).

Slave Quarters

The northwestern corner of the 1710 slave quarters remained a fixed boundary throughout the plantation's history, with no evidence of structures or concentrations of domestic refuse being found further north or immediately west. This may be due to the sizable watercourse running approximately 100 m northwest of this boundary, whose energy may be inferred through the presence of a stone-lined bridge or crossing located where the watercourse and carriage road intersect. The 1755 map better illustrates the size and extent of this watercourse, labeled "Gut," whose cartographic representation suggests that the depth and breadth may have changed frequently.

From this fixed point, archaeological survey of the area identified and documented a series of 55 mounds made up of cut stone, broken window glass, slate, and domestic refuse. Surface debris and ceramics were concentrated within and adjacent to these mounds. Ceramic seriation and glass bottle identification suggest that this area was in use between the seventeenth and twentieth centuries, which is consistent with local accounts of Betty's Hope's continued use up until the 1970s. Stone foundations were partially exposed for some mounds, but ground cover made it difficult to identify structural features for most of the mounds recorded. These mounds were most commonly found in square or rectangular forms; however, there were occasional T-shaped mounds as well. Limited excavations focused on better understanding the nature of these mounds, which resulted in the identification of intact stone foundation walls close to the surface as well as cowrie shells, blue glass beads, and large concentrations of Afro-Antiguan ware. Findings from these excavations are presented in greater detail in subsequent chapters in this book.

Using the 1710 map boundaries for the slave quarters, 17 mounds were identified: 12 rectangular, 4 square, and 1 T-shaped. Using the georeferenced boundaries from the 1710 map and trothing survey results of environmental and structural features, the slave quarters complex measured approximately 3.78 acres of land. The rectangular mounds recorded in this area averaged 8.4 m in length and 4 m in breadth, which is consistent with other accounts of slave house structures in the Eastern Caribbean of the British colonies (Armstrong 1999:179; Edwards 1793:II:126).

While the majority of mounds were found to be between 7 and 9 m in length, they measured as small as 4.8 m and as large as 13.7 m. The width of these mounds ranged almost entirely between 3 and 5 m, with a minimum of 2.9 m and maximum of 6.6 m. All square-shaped mounds were consistently 5.5 m by 4.5 m, while the one T-shaped mound measured 6.8 m, with 4.6 m at its widest breadth.

The 1710 map depicts slave houses as three rows of five structures each, all with their doors facing south. The spatial distribution, along with archaeological excavations and archival research, is inconsistent with this alignment and is more accurately portrayed in the 1755 estate map. When the 55 separate mounds were overlaid on the 1755 slave quarters area, similar spatial configurations were observed. In fact, the area surveyed could account for all slave houses mapped in the 1755 estate plan, each with their associated mound. The primary discrepancy is that additional mounds occur within these boundaries, which may be a byproduct of destruction or structures that may have been added after 1755.

The 1755 boundaries contained an additional 14 mounds: 8 rectangular, 5 square, and 1 T-shaped. These mounds shared similar characteristics as those described in the 1710 area. According to the 1755 estate map, a second slave quarters was settled west of the sugar works. No difference in size or configuration of mounds was observed. The rectangular mounds observed in both areas averaged 7 m long and 3.7 m wide, slightly smaller on average. None of these mounds exceeded 9.5 m in length, while 62.5% were less than 4 m wide. Square-shaped mounds averaged 6.3 by 5.6 m, slightly larger than those found in the 1710 area, while the single T-shaped structure was 9.6 by 6.3 m. The 1755 boundaries for the original slave quarters area to the north increased from 3.8 acres to 6.8 acres, while the added western quarters complex was made up of an additional 2.8 acres.

The mapping and identification of the 1710 slave quarters area was largely the byproduct of destructive bulldozing that occurred in 2014. Bulldozers cleared the vegetation and exposed sheet middens varying across the area northeast of the great house. We were able to take advantage of this unfortunate event to gain access to an area that was previously inaccessible because of heavy vegetation. Similar mounds as previously described were identified to the north (Gonzalez Scollard 2008) and east of the great house containing exposed rocks, cut stone, and occasional bricks. The surface collection of newly exposed deposits found concentrated deposits of Afro-Antiguan ware, broken European wares, and two cobalt-blue glass beads.

An additional 21 mounds were mapped in both the north and western slave quarters areas: 13 rectangular, 5 square, and 3 T-shaped mounds. Rectangular mounds were found to be 8 m long by 4.1 m in width, while square mounds averaged 5.1 by 4.5 m. The three T-shaped mounds averaged 8 m in length and 5.9 m in breadth. The large number of mounds associated with the expanded slave quarters complex is consistent with the almost doubling of the enslaved population and allotted acreage. The enslaved population increased from 227 in 1751 to 393 by 1780, and the original northern slave quarters complex increased from 6.77 acres in 1755 to approximately 10.6 acres post-1755, while the western slave quarters complex increased from 2.8 acres to approximately 4.5 acres.

Although little can be extrapolated from these findings until further excavations have taken place, the variety of mound types may be suggestive of similar spatial variations observed from slave quarters excavations (Armstrong 1999). Excavations are needed to shed light on whether size and spatial clustering can help us understand past social stratification, kinship, and social relations within various periods of enslavement and emancipation. Archaeological investigations have begun to shed light on Afro-Antiguan community planning and architecture.

Provisioning Grounds

Provisioning grounds were critical for the survival and success of both the enslaved and planters alike. The enslaved were given land to grow food for themselves, and planters did not have to use their own capital to purchase costly provisions. Drought, hurricanes, and war meant that imported provisions were unreliable and susceptible to fluctuating prices. In some cases, extreme weather events, such as the hurricane of 1780, resulted in the loss of close to 2,000 slaves on Jamaica and Barbados (Mulcahy 2006:109). Many of these deaths occurred in the weeks following disasters from causes ranging from illness to starvation. Provisioning grounds provided a buffer to unpredictable disasters and economic uncertainties.

Provisioning grounds also provided opportunities, particularly to the enslaved communities. As illustrated in the 1806 painting *Negroes*

Sunday-Market at Antigua and described by Jamaican estate manager William Taylor, Esq., in 1832 in the *Select Committee on the Extinction of Slavery*, the provisioning ground provided an opportunity for slaves to generate surplus commodities to sell at the Sunday markets, which functioned as territory of their own, outside the supervision of overseers and estate managers (Gaspar 1992; Sheridan 1961). However, allotments were not the same as a slave's rights to these provisioning grounds, particularly within the plantation boundaries. As sugar production and labor demands increased, provisioning grounds became increasingly vulnerable to exploitation by the planter class. Betty's Hope provides a good case study into how allotments shifted during a period of economic growth and prosperity for the Codringtons and the island of Antigua.

The quality of land allotted to slaves for provisioning was quite variable; some comprised marginal lands, whereas other allotments were more favorable for food crops (Higman et al. 1998). The provisioning grounds at Betty's Hope in 1710 were located approximately 1.36 km south of the slave quarters along the Tamarin Walk (figure 3.4). These grounds abut a large watercourse to the east and were measured at 47 acres in total. The earliest account for the enslaved population was recorded as 227 slaves in 1751, which results in approximately .21 acres per slave. To put this in context, Lord Seaford's three properties in Jamaica provided 589 acres for 864 slaves, which is equivalent to .74 acres per slave, or 352% more land than at Betty's Hope. Jamaica is considerably larger than Antigua, and further research is necessary to understand how land availability affected the lives of those enslaved. The Shettlewood Estate provided 1,200 acres for 421 slaves, which is equivalent to 2.78 acres each or an astounding 1,300% greater (Higman et al. 1998; Mullins 1995). It is important to point out that Betty's Hope experienced two periods of expansion—one in 1737 when the north windmill was added and another in 1741 during the addition of five copper boiling vats—and that there may have been significantly fewer slaves when the estate was surveyed in 1710 with respect to their allotted provisioning grounds (see figure 3.4).

By 1755, the provisioning grounds at Betty's Hope were converted to pastures for the procurement of manure necessary to maintain soil fertility. Suzanna Pratt's (2015) master's thesis used the USDA's erosion-productivity impact calculator to simulate the impact of intensive sugar monoculture against actual sugar yields and found that Betty's Hope continued to outgrow the simulation, suggesting that management and land-

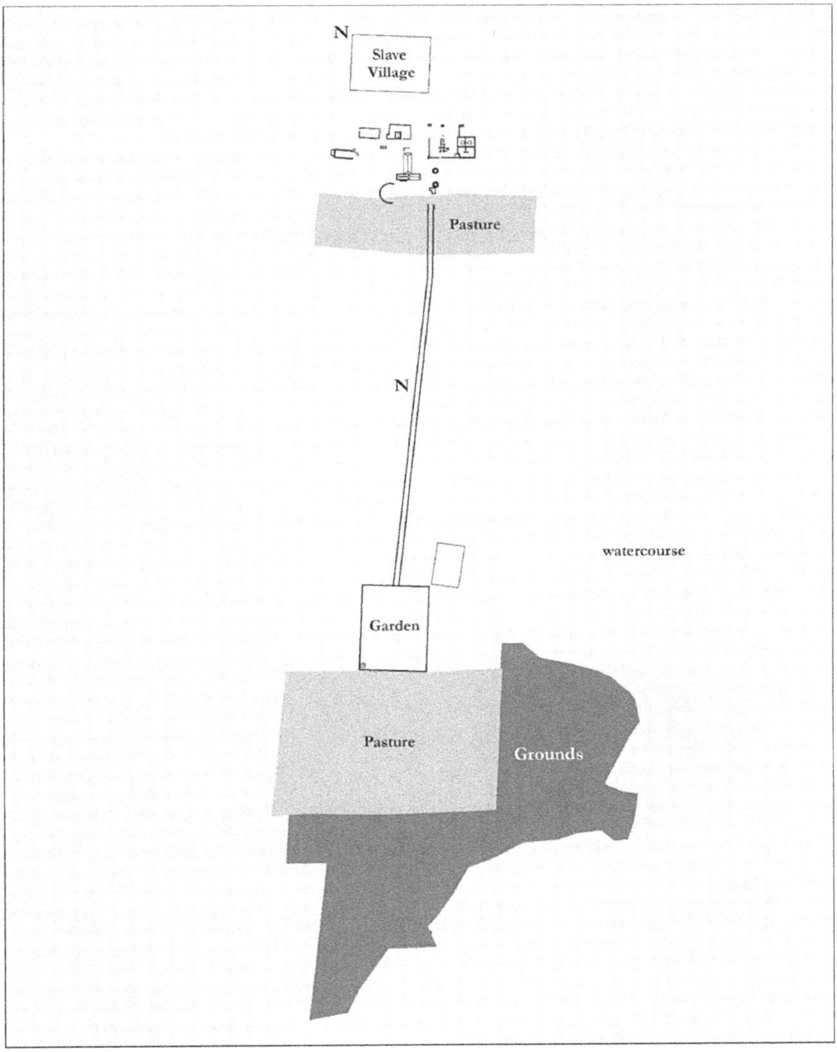

Figure 3.4. Map illustrating how space was reorganized to suit the growing sugar production at Betty's Hope. The 1710 provisioning grounds for the enslaved were turned into pasturelands, primarily for production of fertilizers and as grazing land for cattle.

use practices may have contributed to the farm's ongoing success (Pratt 2015). From 1707 to 1753, the plantation experienced its largest period of growth in sugar yields, while from 1710 to 1755, the lands used for pastures increased from 31 acres to 120 acres. The planter's decision to shift the original pastures, located just south of the garden, to the provisioning grounds resulted in a net gain of 89 acres for grazing and manure production. This net gain, however, translated into significant losses in provisioning grounds by 1755. The 1755 estate map neither demarcates nor lists acreage for provisioning grounds. It is more likely that enslaved Africans relied more on the small gardens in the slave quarters as the provisioning ground acreage disappeared.

In fact, it becomes abundantly clear that lands surrounding the Codringtons' five estates were completely utilized for agricultural production. If pastures were restricted by planters for provisioning, then these grounds would have been concentrated around the boundaries of each of the five estate-related slave quarters. While acreage was not recorded for these areas, georeferencing and ground truthing of the 1755 estate map provided an estimate of 29 acres for all five slave quarters. Using the 1751 records of those enslaved provides a rough approximation of .13 acres of land per slave, or a 38% decrease in land used for provisioning.

By 1780, the enslaved population was recorded at 393, making only .07 acres of land available for provisioning, or a decrease of 67%. Survey of the post-1755 slave quarters area surrounding Betty's Hope showed that the slave quarters increased by only 5.5 acres, which still appears woefully small when compared with other plantations. Barbuda, Antigua's sister island to the northeast, which was granted to the Codrington family in 1684 and continued for the next 200 years (Lowe 1951:44), may have provided some support for slave provisions. Taking this into account, the loss of provisioning grounds for those enslaved at Betty's Hope was substantial.

This raises the question of what impact the economic success of Betty's Hope had on those enslaved. The loss of provisioning grounds would have had a significant impact on those producing for the Sunday markets and would have undoubtedly created new tensions regarding the territoriality of land surrounding slave houses. Provisioning grounds created meeting places outside the watchful eye of those in power where information and ideas could flow throughout different quarters. It is possible that managers and overseers viewed these spaces as threats to their safety, while heavy economic tolls played on the minds of the planters. As recorded in 1736, an attempted rebellion resulted in the subsequent year's heavy losses (Pratt 2015), although the English sugar market had already been experiencing a recession throughout the 1730s, making matters worse (Gaspar 1978:316). The disappearance of dedicated provisioning grounds coincided with a time of growth and prosperity, a time when the growing slave population was being divided among five estates instead of two.

How these reconfigurations of land reflected a slave's overall mobility is less certain, as the 1736 plot to attack the island's elite in the capital town of St. John's was a coordinated event that included slaves from across the island (Gaspar 1978). Robert Arbuthnot, a justice of the peace, described the commonality of slaves meeting around church grounds and pastures, only raising suspicion when more frequent meetings were observed after midnight (Gaspar 1985:21).

St. John's was a common place for those enslaved to meet on Sunday market days at one of the many grog shops and dram shops scattered throughout the town. Rumors in 1831 that the Sunday markets might be abolished resulted in island-wide anxieties and fears over the possibility of slave uprisings and revolts (Gaspar 1988, 1992:138). Places where slaves could gather and exchange ideas shifted power in ways that threatened the lives of managers and overseers, despite the plantation's layout and topography. Greater research is needed to understand how these spaces helped shape resistance and the emergence of an African Antiguan identity.

Land-Use and Landscape Change

As other chapters in this book will go into greater detail regarding the buildings and structures associated with planter landscapes, this section focuses more on the structural fluidity and land-use changes related to sugar production. As previously discussed, buildings directly associated with the production of sugar were constantly modified, repurposed, and built anew, making them poor references for situating centuries-old estate maps. In fact, the southern windmill represents the only structure that can be confidently sourced to the 1710 estate map, which experienced minimal modification. The reliance on situating planter spaces falls again to the boundaries and walking paths; however, these same boundaries pose different obstacles and create different meanings for those in charge of managing and running the estate.

Increasing Productivity

Betty's Hope experienced three major expansion phases from 1710 to 1827. However, the period between 1710 and 1755 represents a formative shift, with two plantations located within 1,323 acres of land to five plantations in 1,426 acres of land. While the total acreage increased by 103 acres, or 7%, the majority of land became increasingly subdivided for sugar monoculture. For instance, while the original boundaries for many of the "sugar pieces" or land parcels were preserved, these parcels became subdivided into smaller plots during this time. The difficulty in gauging the relative production and sugar yields for Betty's Hope is tied to these new land-use strategies, with 871 acres making up the plantation in 1710, changing to 543 acres by 1755. The nearby Codrington Cotton Estate also shrank from 871 acres to 543 acres during the same time period. The increase in plantations associated with smaller parcels of land may have allowed managers to increase the efficiency and cultivate the land more effectively. During 1710, a total of 703 acres, or 81%, of the total land at Betty's Hope was used for sugar production, while "sugar pieces" shrank to 477 acres in 1755, thereby increasing sugar production to 88% of the total land.

Six windmills would have given the Codringtons a great deal of production capacity for their acreage, and as captured in Herman Moll's 1736 plantation map of Antigua, Betty's Hope was a "Plantation of Great Note." This map labels Betty's Hope separately, despite the fact that its twin windmills are included under a separate label for Codrington's six windmills. The Herman Moll map also captures one of the expansion periods, providing a time frame for when the other windmills were constructed. As previously stated, the north windmill was added in 1737, a year after this map was produced.

It should also be noted that neither of the two archival maps demarcates the presence of a forest, which was often included on other estate maps as a result of its intrinsic value and aesthetic appeal (Buisseret 1996, 2010; Higman 2001). Betty's Hope appears to be consistent with the intensive land clearing and deforestation that took place between 1690 and 1751, with the removal of two-thirds of Antigua's forests for sugar monoculture (Watts 1987:434–435), which, it seems, Betty's Hope was not immune to. Mangroves are listed for the Cotton Plantation in 1710, although their labels are nowhere to be found. What makes this particularly interesting is that the closest modern mangrove system is 1.3 km to the northeast of Betty's Hope along Mercer's Creek, outside the range of the Codringtons' property. This suggests that either the mangroves were included in the legend for accounting purposes and were not present within the Codrington estates or that some degree of environmental change has occurred over the past two centuries.

The closest body of water is the watercourse running 700 m south of the Cotton Estate bordering Betty's Hope. Regardless, the relative lack of mangroves and intact forests as noted by estate surveys raises the question of what kinds of building materials were available for the enslaved. Wattle houses do not preserve well, and those that do are often in poor condition. While plantations such as New Montpelier did not see the advent of stone houses until 1825 (Higman et al. 1998), the lack of natural resources available for construction suggests that stone structures may have been an early part of slave architecture at Betty's Hope (see chapter 12 in this volume). Continued archaeological investigations of the estate's slave quarters are necessary in order to understand the relationship between available resources and the early architecture of these structures.

Colonial Adaptations to Drought

Drought and water shortages were a constant concern for planters at Betty's Hope, as evidenced by the four deep cisterns added by 1755 and the construction of another large cistern post-1755. It has been argued that water resources played a significant role in the manner that plantations were organized (Higman et al. 1998); however, Betty's Hope demonstrates the role that social-ecological-technological systems played in building a resilient infrastructure for sugar production (Fox 2016). In contrast to plantations like Montpelier in Jamaica, with its numerous rivers flowing through the sugar-works complex, Betty's Hope relied solely on a mix of dug wells or ponds and massive cisterns. The only natural body of water was the brackish watercourse flowing along the northwestern boundary of the site, crossing north of the slave quarters and continuing west. Thirteen ponds or shallow wells were demarcated on the 1755 estate map across the five Codrington estates, of which ten are still in use today. Some collection ponds or wells were also stone lined, such as the one at Betty's Hope directly south of the sugar works, which still retained a portion of a cut stone wall aligning its perimeter. Betty's Hope consisted of two ponds, a massive cistern with its four separate tanks measuring 35 m by 40 m, and a post-1755 addition of another cistern measuring 10 m by 6.7 m (see figure 3.5). This expansion was described in the Codrington Papers housed

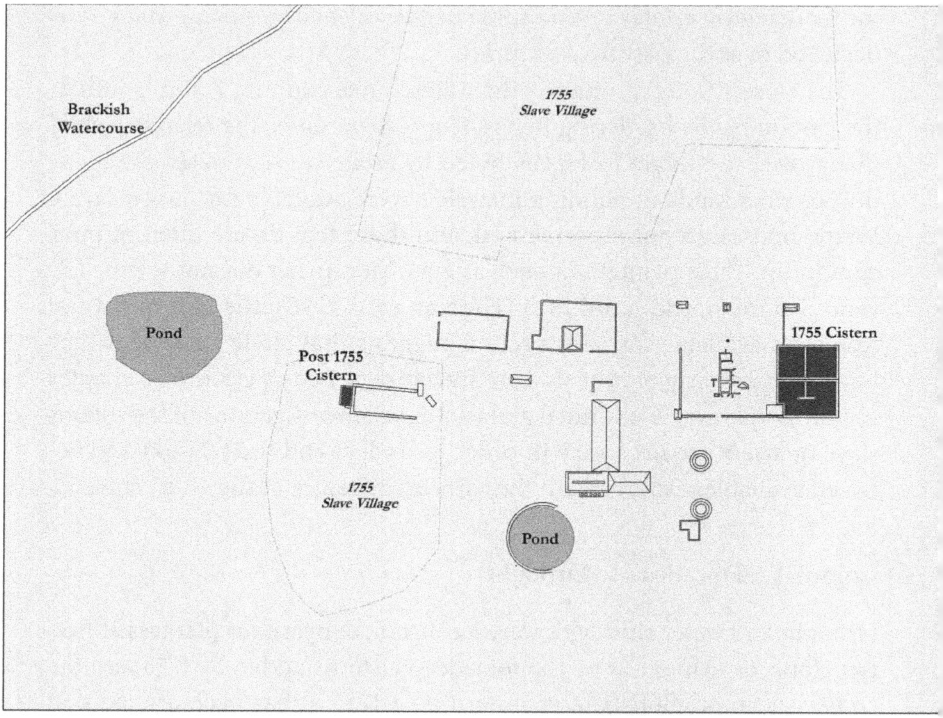

Figure 3.5. The extensive water management system at Betty's Hope is consistent with the 1790s description of technological expansion of the estate's water holding capacity, consisting of two ponds, a massive cistern with four separate tanks, and an additional cistern.

at the National Archives of Antigua and Barbuda, where correspondence described the expansion of cisterns that was completed by the 1790s.

After the section north of the great house was bulldozed, two additional circular depressions were observed, both similar in size to the adjacent ponds, suggesting that post-1755 modifications to the landscape may have been added to meet the demands of sugar monoculture. The continued expansion of the estate's water-holding capacity may have played a role in maintaining sugar production despite extensive periods of drought. During the 1780s and again in the 1830s, Antigua experienced heavy periods of drought; however, Betty's Hope's sugar production continued to keep within the estate's six-year average (Berland et al. 2013; Pratt 2015).

Since the seventeenth century, water management has been a part of the estate's history. The continued construction and modification of the landscape was a combination of small incremental additions, such as ponds and wells, and large-scale constructions, such as the massive cisterns sunk into the ground. Without these short- and long-term adaptations, far more droughts would have been recorded along with a different trajectory for the plantation's success.

Reconstructing Spaces of Tension

Betty's Hope was in many ways a model for economic and environmental sustainability in the sugar industry. With a small amount of land in an arid place, the Codringtons produced one of the most successful sugar plantations in the Lesser Antilles. Evidence suggests that planters and managers of the estate integrated short-term incremental adaptations mixed with long-term reorganization of space to mitigate the impacts of drought and soil exhaustion. The decisions to increase the estate's resilience to these conditions exploited the people who were most socially vulnerable. Early land-use decisions, particularly between 1710 and 1755, altered the trajectory for those laboring on these lands as well as the landscape legacies inherited by modern farmers today.

What is inscribed in the history of spaces at Betty's Hope are environments and boundaries that were constructed by generations of planters, laborers, and the enslaved. Their relationship to nature, particularly how it was accounted for and placed in estate ledgers, sheds light on the economic value of managing soils and manure correctly. With an increase of pasturelands by almost 400%, the production of manure seemed to help offset the degradation of soils that was commonly observed throughout Jamaica, Barbados, and other plantations on Antigua (Higman et al. 1998; Sheridan 1960). Further research is needed, particularly in the field of bioarchaeology and environmental archaeology, to understand the full impact that land-use decisions may have had on the inhabitants health, disease, and diet.

The demand for sugar during the eighteenth century helped to socially construct the meaning of drought for much of the West Indies. The expansion of Betty's Hope's water-holding capacity meant that planters could dampen the impact of dry years on a six-year scale and sustain a stable and predictable yield for the land. Descriptions of drought were therefore always being recalibrated depending on the holding capacity of the estate. Certainly, drought years did exist, but technological modifications to the landscape helped redefine the problem of water as a natural phenomenon and converted it to a problem of technological solutions. Similar issues are plaguing Antigua today. The increase in desalinization plants has helped mitigate the ongoing impacts of drought on the island; this growth in technological solutions has coincided with a period of increased development in industries related to water consumption, particularly a renewed focus on agricultural endeavors. Requests were made by Antigua's Environment Division for case studies that demonstrate how early land-use and management decisions can affect the trajectories of both its land and its people. Case studies such as this can have broader impacts by contributing locally toward stakeholder and practitioner knowledge production, providing places for people to see, revisit, and experience.

PART II

Maintaining the Plantation

4

Barbuda and the Provisioning of the Codrington Estates on Antigua

JENNIFER L. ANDERSON

In December 1762, Samuel Redhead, manager of William Codrington II's Antiguan plantations, including Betty's Hope, wrote to his employer bewailing a terrible drought that was destroying the sugar crop and exacerbating a distemper that killed many cattle. "I should have lost a great many more if I had not sent them to Barbuda where we had grass enough for them," he reported. "We should have been vastly distressed if this Island was not in our possession at this time" (Redhead, December 12, 1762). Convinced of the tiny island's importance to his employer's interests, Redhead expressed frustration that Codrington, an absentee owner residing in England, seemed not to fully appreciate its usefulness. Barbuda was a sore spot for Redhead because until recently, it had been subleased for a fifteen-year term (1746–1761) to Samuel Martin and William Byam, who had profited considerably from the arrangement. When their sublease expired, Redhead opposed seeking another tenant, insisting that any rent received was worth less than the island's produce. Accounts of the increased value of livestock and enslaved Africans during Martin and Byam's tenure bolstered his case, likely weighing into Codrington's decision to retain Barbuda, placing it under Redhead's supervision. Nevertheless, Codrington expressed perennial dissatisfaction with the island's underdevelopment and regularly demanded improvements to boost its productivity.

Despite Codrington's unfulfilled expectations, his family's exclusive control over Barbuda proved beneficial for them throughout the eighteenth century—apart from when Martin and Byam reaped its benefits. Although long dismissed as unsuitable for sugar cultivation, Barbuda helped the Codringtons' Antiguan plantations prosper and overcome times of adversity that bankrupted other proprietors. Over many decades, Barbudan produce, coaxed forth from arid land and meager natural resources, sustained the enslaved workers and sugar operations at Betty's Hope. For enslaved Africans dispatched to Barbuda, its atypical labor structure allowed them relative autonomy and flexibility, much preferable to the relentless toil and oppressive supervision in cane fields or sugar mills. While sometimes overstated, historical accounts have often characterized Barbuda as a much healthier locale than Antigua. Yet the difference was significant enough that many black Barbudans disliked Antigua, so much so that being sent there was used by their masters as a form of punishment. Moreover, individuals accustomed to life on Barbuda often fared poorly when impressed into arduous labor at Betty's Hope. Miserable to be separated from friends and families, they begged to return to their home island (Lightfoot 2015:34). Banishment to Antigua proved a death sentence for many who never saw their loved ones again. Even so, Barbuda had its own problems—ranging from negligent or abusive managers to taxing environmental conditions—that caused hardship for its inhabitants, especially the majority enslaved population. Like most of the Antilles, the island was also vulnerable to droughts and hurricanes, although their impact tended to be worse on Antigua as a result of the ecological vulnerability of its sugar monoculture (Berland et al. 2013; Pratt 2015).

To understand the vital supporting role Barbuda played for Betty's Hope, one must consider two main factors: first, what it contributed to the plantation complex, serving as a valuable ancillary to sugar growing; and second, how its particular labor regime and ecological conditions were experienced by the people consigned to live and work there. On balance, the fact that Barbuda was never converted to sugar had far-reaching consequences both for its essentially feudal status within the Codringtons' West Indian empire, and for the enslaved people who upon emancipation ultimately inherited the island.

During the early years of European colonization, prospects for Barbuda did not seem promising. As sugar cultivation took off elsewhere in the Caribbean, assessments of the tiny island (161 km^2)—located 50 km northeast of Antigua and about half its size—deemed it too small, dry, and barren for growing cane. By the late seventeenth century, only a handful of hardscrabble English farmers had immigrated there. In 1684, the English Crown granted a long-term lease for the island to Antiguan planters Christopher and John Codrington. While Christopher focused on Betty's Hope, his brother took up the task of developing their new leasehold. Concentrating on low-cost ventures, John dedicated part of Barbuda to provision grounds and the remainder to free-range grazing (Parker 2011:166–67, 210; Sluyter 2012:103).

The brothers' sole access to Barbuda curbed one of the greatest economic strains faced by West Indian planters, namely the high cost of imported supplies. As discussed in the previous chapter, Antigua, like most sugar islands, relied predominantly on comestibles, wood, and other essential materials shipped in from North America, Ireland, and England. With Antigua's best land reserved for sugar, provisions grounds and garden plots were relegated to marginal areas, and enslaved workers were permitted to tend them only during their limited personal time. Local food production thus remained insufficient for the whole population, even though it was cost effective and provided essential nourishment. So long as sugar prices remained high, Antiguan planters prioritized growing cane, even if that meant spending more on imported provisions—dried, salted, or brined—that were more expensive and less wholesome than fresh local foods. Access to nearby Barbuda was thus extremely advantageous for the Codringtons as they could oversee production, slash transatlantic shipping costs, and eliminate middlemen. Barbuda's importance to their operations only increased as environmental conditions on Antigua deteriorated.

By the late seventeenth century, the Codringtons, Byams, Martins (owners of neighboring Greencastle), and a few other families emerged as a powerful plantocracy on Antigua. The settler population then included approximately 3,000 white inhabitants and 10,000 Africans, whose subordinate position in the racial hierarchy was enforced by draconian slave codes (Dyde 2000:41). As Antigua's first flush of productivity outpaced other British sugar islands, proprietors enjoyed enormous profits. The island's singular focus on sugar, however, resulted in serious ecological problems, including extensive deforestation, soil compaction, erosion, and flooding. Growing cane also depleted the soil after a few years, so fertilizing wasted fields became essential (Dunn 1972:30; Richards 2003:426, 460; Richardson 1992:29–32). Most challenging was the island's dearth of freshwater, worsened by constant desiccating winds, unpredictable rainfall, and a lack of dependable streams or springs. Although plantations were engineered to capture as much rain as possible in ponds and cisterns, water shortages remained a chronic, at times life-threatening, problem. Recurrent droughts also plagued the island, lasting for months or years, necessitating emergency forays to fetch water from other islands (Dyde 2000:5).

Although Antiguan born, William Codrington I, who inherited the bulk of his family's West Indian holdings in 1710, opted to move to England rather than have his children endure such privations. Among the property he left behind were the enslaved people on Barbuda, who numbered 92 in 1719, as well as 20 white indentured servants who still owed him labor. As a member of the peerage, Codrington secured a place in Parliament but continued to keep an eye on his plantation holdings from the comfort of Dodington Park, his family's ancestral seat (Parker 2011:205). Upon his death in 1738, his eldest son, William II, who was only 19 and had never left England, took ownership of the Antiguan estates and the lease for Barbuda. Given that Antigua was still recovering from a major social upheaval, however, there was no question of the young heir moving there.

Two years earlier, plans for a purported slave revolt were uncovered on Antigua, prompting a ruthless response from planters and colonial officials. Although historians disagree about how serious such a plot may have been, conditions certainly were ripe for an uprising. Amidst a terrible drought, the captive population was famished and desperate. Rumors of slave unrest swirled, causing many to dread the worst. According to planter Josiah Martin, the "intended Insurrection of Negroes [aimed] to destroy all the white Inhabitants of this Island, without Mercy or Distinction, [but] by the good providence of God was discovered." Colonial officials arrested and tried hundreds of suspected participants, using torture to extract confessions. "Many Negroes [were] put to death different ways, some hung up alive in Gibbets, some burnt, and others broke over the Wheel" (NAAB 1737; Martin, December 10, 1736). Three men owned by Codrington—Sacky, Jacko, and Ghlode—were among those gruesomely executed. Before the panic subsided, 88 people lost their lives, and 44 were deported (Gaspar 1985:30–34; Mulcahy 2006:78; Sharples 2012:35–59). In the aftermath, many prominent planters left the island and absentee ownership became the norm. William Codrington likewise entrusted his Antiguan holdings to hired surrogates rather than move there. The recent trauma on the island likely highlighted for him and his advisors the volatility of life in a full-blown slave society.

Barbuda's Contribution to the Plantation Complex

By the mid-eighteenth century, Betty's Hope had grown into a substantial estate with a large enslaved population. In 1754, William Codrington owned "963 Negroes, old and young," of whom 799 were on Antigua and 164 on Barbuda. At that point, Redhead estimated that since nine had recently died from overexertion, an additional 150 was needed to bring the total workforce to "a pretty complete strength and enable us to do great things" (Redhead, June 8, 1754). Betty's Hope must have seemed like a small city. With so many mouths to feed, providing adequate sustenance demanded careful planning and considerable resources. In addition to imported rations of beans, grains, and other basics, plus whatever residents could grow or forage, Betty's Hope now received regular deliveries of fresh food from nearby Barbuda. With enslaved Africans providing the labor, the island produced vegetables (including yams, maize, and potatoes), fish, and meat (including beef, pork, mutton, and goat)—although most of the meat was probably reserved for Antigua's white population. Any surplus was sold to neighboring plantations, generating added revenues for Codrington. Barbuda also supplied Antigua with other necessary commodities, such as leather, salt, lime, and wood for fuel and fencing. Access to this satellite location thus gave the proprietor of Betty's Hope a decided edge over planters who remained heavily dependent on imported provisions.

Barbuda's sparsely populated shrubland proved especially suitable for ranching. Over the years, the islanders formulated their own system of animal husbandry that combined free-range grazing, herding, and the use of cattle pens (Sluyter 2012:98–139). Cattle wandered through brush thickets in search of forage and watering spots. Periodically, the enslaved herdsmen rounded up the feral animals to be penned and fattened prior to shipment to Antigua. In addition, they raised horses and asses for riding and pulling carts. Flocks of sheep and goats also dotted the hillsides. The need for meat and draft animals on Antigua created steady demand for Barbudan livestock. Codrington even secured lucrative contracts to supply meat to British naval vessels stationed there. Large quantities of manure, a valuable byproduct of Barbuda's stock-keeping operations, were also shipped to Antigua to serve as fertilizer.

In a creative adaptation to local conditions, Barbudans constructed corrals that enclosed masonry wells, thereby constructing watering holes that they could refill as needed with buckets or pumps. During the dry season, thirsty cattle gathered within these walled oases where they could be contained until needed. The enslaved workforce included able-bodied men, who served as herdsmen and were responsible for the livestock. Women, children, and less fit or elderly men also labored, cultivating the provision grounds and gardens, fetching water, gathering wood, and cooking meals, among myriad other tasks. Since many of the first enslaved people on Barbuda originated in cattle-raising regions of Africa, they may well have had prior experience in keeping livestock. Historian Andrew Sluyter's comparative analysis of ranching in the Americas confirms that Barbudans incorporated Senegambian, English, and Spanish Caribbean herding practices, including herd dogs, horses, lassos (cast on foot and from horseback), and tail twisting (a technique used to bring down cattle). The construction of wall-enclosed wells, however, appears to have been "unique to Barbuda and presumably represents a radical local innovation" (Sluyter 2012:135).

Efforts to introduce other kinds of economic activities on Barbuda proved less successful, although Codrington directed his managers to experiment with numerous commercial crops. At various points, for example, attempts were made to grow opuntia (a type of cactus used as fodder), aloes, cotton, and corn. Codrington also toyed with introducing the cultivation of coffee and East Indian spices, but apparently never implemented the idea. While some substantial amounts of cotton were produced, the labor-intensive task of cleaning the ripe bolls—before the advent of the cotton gin—made it unremunerative. Except when a lack of rain stunted its growth, Guinea corn thrived well and, in ample years, served as the main dietary staple for Codrington's workforce on both islands. Since Barbuda also provided fodder for his livestock, a reliable supply of hay and other provender was essential. To supplement the island's natural grasslands, Redhead ordered the seeding of acres of Guinea grass (*Panicum maximum*), a species of African grass inadvertently introduced to the Caribbean in the early colonial period. Prolific, drought tolerant, and nutritious, the desirable grass soon was transplanted throughout the region (Watts 1987:428; Galloway 2005:104). After one harrowing dry spell, Redhead expressed relief that thanks to "the propagating of Guinea Grass," only one old cow had died (Redhead, March 21, 1771).

Challenges of Living and Working on Barbuda

After Samuel Redhead initially took charge of Betty's Hope circa 1751, he updated Codrington regularly about conditions on the plantation and prospects for the sugar harvest (Lowe 1951). At the time, Martin and Byam still had 10 years left on their lease agreement, which included responsibility for all the Codrington-owned people and livestock on Barbuda, in addition to any of their own. Codrington likely had been motivated to off-load Barbuda to them, at least temporarily, after a slave uprising occurred there in 1745. Apparently unbeknownst to him, the manager at the time was so cruel and brutish that his victims eventually revolted and murdered him. Their distress was further magnified by hunger caused by a severe drought. From Antigua as well, reports came that "the Cattle are starving and begin to die" and there are "Little or no provisions in the Country to feed White or Black" (*London Evening Post*, November 29, 1746). Nevertheless, colonial officials on Antigua quickly mustered military reinforcements to restore order on Barbuda. This disturbing incident no doubt harkened back to the shocking tales of violence and vengeance after the purported slave insurrection of 1736. Facing a similar situation on Barbuda, Codrington seems to have concluded that the meager island, seemingly ungovernable and marginally profitable, warranted little further investment.

Codrington's negative view of Barbuda would change, however, as it became evident that where the murdered manager abysmally failed, Samuel Martin and William Byam succeeded, transforming Barbuda into a well-run, lucrative stock-keeping operation. Their involvement stemmed from Martin's efforts to revive Greencastle, the Antiguan plantation he inherited from his father, which had deteriorated after a long period of absentee ownership. Martin took over its management himself, upgrading its equipment and investing in additional labor. But its depleted soil and the high cost of provisions undermined his efforts. Recognizing Barbuda's potential to alleviate both issues, he and his business partner, William Byam, thus offered to take it off Codrington's hands. As better food and more fertilizer began to arrive from the island, Greencastle rebounded. Martin and Byam ensured that the people and beasts on Barbuda regained prime condition as well. During the term of their lease, the value of the partners' stock—human beings and animals—increased by over £18,000 (Tweedy 1981:66).

Based on these experiences, Samuel Martin later published a popular planter's guide that offered guidance on sugar cultivation and plantation management. Significantly, he also advocated for improved living and working conditions for slaves. When Codrington requested a copy, Redhead promised, "I have not been able to meet with Col. Martin's *Treatise on Planting*, as soon as I can, I will send it to you" (Redhead, May 3, 1755). Since Martin tempered his embrace of slave-powered capitalism with a degree of paternalism, historians have often characterized him as an early proponent of amelioration (Sheridan 1960:126–139; Zacek 2007:8–31). His motives, however, were more profit oriented than humanitarian. He argued, for example, that a master who "feeds his negroes well, proportioneth their labour to their age, sex, and strength, and treats them with kindness and good nature, will reap a much larger product, and with infinitely more ease and self-satisfaction than the most cruel Egyptian taskmaster, who starves his negroes, or chastises them with undue severity" (Martin as cited in Rebovich 2015:68).

Eyeing Martin's success with envy, Redhead convinced Codrington that it was "absolutely necessary for the carrying on of your estates to hold at least half of [Barbuda] yourself" (Redhead, July 18, 1758). In 1761, Codrington, in fact, resumed control of the whole island (which his descendants retained for another hundred years), appreciating now more fully how it complemented his Antiguan holdings. Emulating Martin and Byam's approach proved difficult, however, because Codrington could not find a qualified manager who had the requisite knowledge and experience and who was willing to move to such a bleak location. In 1771, Redhead relocated to Barbuda instead, ostensibly to supervise operations there himself. In truth, he wanted to be closer to Sarah Bullock, an enslaved woman whose freedom he secured from Codrington. Emboldened by her influence over the doddering manager, the Bullock family gained a mafia-like dominance on the island, overseeing their own affairs, spurning required work assignments, and wielding considerable influence over the rest of the enslaved community.

After Redhead retired to England (accompanied by Sarah), Dennis Reynolds took up the vacated post and proved a much more competent manager than his predecessor. Serving from 1779 until circa 1793, he diversified Barbuda's agriculture and increased its livestock. One of his first lines of attack, however, was to discourage what he regarded as poaching. Long accustomed to helping themselves to the animals they tended every day, the enslaved community had come to regard unlimited fresh meat as an informal right. For years, managers and overseers failed to curb the practice. By one account, the 1745 slave revolt was triggered in protest when some herdsmen were penalized for butchering cattle without permission (Tweedy 1981:191).

Regardless, Reynolds felt discipline was lacking. In his view, if workers were "properly employed in doing their duty and made industrious in [cultivating] their grounds on Saturday afternoons and Sundays . . . they could not have had the opportunity of destroying so many [animals]." Accusing the Bullocks of being ringleaders, who "were so insolent as to tell me they never would be hungry while there was a Cattle on the Island" (Reynolds, May 22, 1792), he banished the culprits to Antigua. When Codrington complained of "the refractory disposition of the Negroes," his attorney, Langford Lovell, blamed Redhead, who "ruined them with Indulgence and they were then in the habit of killing Sheep . . . and severe examples have been since tried to break them from it" (Lovell, January 18, 1792). Undeterred by their master's resentment, they continued to enjoy steaks and spare ribs at his expense.

Barbudans were not immune, however, to larger environmental challenges that affected the entire region, especially natural disasters. Droughts, stretching over months or years, proved particularly devastating, destroying crops and livestock and causing considerable human suffering. At other times, hurricanes wreaked sudden havoc, wiping out fields, provision grounds, freshwater supplies, and food stores (Mulcahy 2006). When natural disasters struck, everyone suffered. But while planters could always just leave until the emergency passed, the captive workforce did not have that choice. As occurred amidst the drought and resulting famine in 1736 and again in 1745, people who felt trapped, with little left to lose, could be driven to revolt. Colonial officials and planters, in turn, resorted to increased repression to dampen the social powder keg and quell their own fears of slave uprisings.

At least up to a point, Barbuda usually weathered dry spells better than Antigua with its larger concentrations of people and draft animals. In March 1772, for example, Redhead reported that Antigua was "in great distress for want of water for the stock and if we are not soon relieved the mortality will be great." Although the dwindling springs had not yet disappeared, he feared thirst-wracked residents would be tempted to drink tainted water. Meanwhile, Barbuda still "flourisheth very well, but it is now very dry. The Stock as yet keeps up tolerably" (Redhead, March 6, 1772). After another hurricane the following year, Redhead again observed that Antigua suffered "infinitely more from the Droughts" (Redhead, May 25, 1773).

In 1790, Langford Lovell informed Codrington that a lack of rain once again threatened both Antigua's sugar crop and Barbuda's corn, so critical "for the support of the Negroes." As the drought dragged into its second year, he reported a "great loss of cattle and other stock . . . when Food for their Subsistence was withheld by the hand of Providence" (Lovell, November 16, 1790, and January 18, 1792). At Betty's Hope, they resorted to feeding brittle cane tops to the cattle but still lost more than on Barbuda. Fortunately, the rains came just in time. To everyone's relief, soon the "Negro grounds were full of Provisions and there was a promising appearance of Guinea Corn which makes any further Supply of Provisions from [England] unnecessary" (Lovell, January 18, 1792).

Except in times of crisis, the people enslaved on Barbuda generally were stronger, healthier, and more disease resistant than their Antiguan counterparts, thanks in larger measure to their varied diet, including adequate protein, and less harsh working conditions. Their rates of natural increase were also higher than West Indian averages (Berleant-Schiller et al. 1995:52; Higman 1984:348; Lightfoot 2015:34; Lowenthal and Clarke 1977:510–535). In the perverted logic of the slavocracy, of course, their physical well-being also enhanced their value as property, more than compensating their master's bottom line for all cattle they consumed. When William Codrington died in 1792, his nephew Sir Bethell Codrington took possession of Barbuda and its robust inhabitants. Upon emancipation in 1834, he had the temerity to demand extra compensation, claiming that his human chattel were especially valuable because of the island's "extreme fitness," a request that was denied (Lightfoot 2015:34).

By then, the people of Barbuda had already cemented their reputation for stubborn independence. In 1829, for example, a manager who left after a brief stint complained that "they acknowledge no Master and believe the Island belongs to themselves, and any manager living [in Barbuda] & using coercive means to bring them into subjection I have no doubt would lose his life" (cited in Lightfoot 2015:35). In the wake of emancipation, the Codringtons largely neglected the island while denying its inhabitants both employment and access to land. Consequently, Barbudans "endured a protracted and difficult transition to freedom because their labor was no longer profitable to the white proprietors" (Lightfoot 2015:204). Nevertheless, many chose to stay on Barbuda, expressing a strong affinity with their home whatever its drawbacks. After generations of toiling in bondage for the Codrington family—driving wild steers across the rugged landscape, coaxing crops from the unforgiving soil, and creating a unique way of life—the freed people of Barbuda had indeed made the island their own.

Conclusion

During the colonial period, Barbuda's built landscape, modes of land use, and labor system were all geared toward supplying the needs of Antigua's sugar plantations. With the decline of the sugar economy in the nineteenth century, however, its specialized economy was no longer viable. Since the island was never subdivided under the Codringtons' sole proprietorship, freed Barbudans established their own form of communal land tenure that allowed for the continuation of free-range ranching and subsistence farming (Potter 2015). With few natural resources and little of value to export, the island remained largely excluded from international markets. Facing widespread unemployment, many people were forced to leave the island in search of new opportunities, including relocating to Antigua, where the 1834 Contract Act forced its freed people to work for their former masters. Those who stayed on Barbuda eked out modest livelihoods but enjoyed considerable autonomy, self-sufficiency, and a strong sense of shared Creole identity (Berleant-Schiller 1977:253–72; Lowenthal and Clarke 1979). Consequently, Barbuda developed "a life of its own," making it hard to categorize within the larger West Indian context (Higman 1984:66).

Although now politically united with Antigua, the Barbudan people still take great pride in their history of resilience and their distinctive culture. Yet over the last decade, the island has continued to face environmental challenges, especially water shortages, and contentious disputes over changing land uses and "the vexed question of land tenure" (Boger et al. 2014:331–339; Phillips 2013, 233–236). Efforts to attract tourist dollars to this less-trodden destination have been accompanied by growing external pressure to build new hotels, resorts, roads, and other supporting infrastructure. To stem that tide, the Barbuda Land Act of 2007 codified communal land tenure into law and effectively limited private development (Potter 2015:110–113). Land may be leased by private enterprises for up to 50 years, but only if approved in a public referendum by a majority of Barbudans (Gould and Lewis 2018:148).

After generations of survival, self-determination, and relative autonomy, however, the Barbudan community and way of life are now under dire threat. When Hurricane Irma smashed into the island on September 6, 2017, it wrought a level of destruction never witnessed there in modern times. The entire population was evacuated, and almost every structure was damaged or ruined. In the immediate aftermath of this apocalypse, Ronald Sanders, Antigua and Barbuda's ambassador to the United States, expressed dismay that for "the first time in 300 years there's not a single living person on the island of Barbuda" and its civilization "has now been extinguished" (deGrandpre 2017). While this claim of cultural annihilation, made in the immediate aftermath of the storm, hopefully was premature, the economic devastation can hardly be overstated. Many Barbudans remain dispersed in temporary accommodations on Antigua. Those who have managed to return are struggling to rebuild.

At the same time, outside developers and regional proponents of international investment have rushed to take advantage of the emergency in hopes of snatching up newly available real estate. Such efforts constitute a new form of economic colonialism termed "disaster capitalism," wherein governments and private interests exploit national catastrophes to push through controversial policies, such as privatization of communal lands, while citizens are too stressed or distracted to resist (George 2017; Klein 2007:6). Despite opposition from the islanders themselves, the national government of Antigua and Barbuda, under the leadership of prime minister Gaston Browne, has been receptive to such overtures and seems poised to strip Barbudans of their common land rights. In a stunning feat of historical revisionism, Browne has gone so far as to reject the very concept of communal landownership as "a fallacy, it's a myth, it never existed." His message to displaced Barbudans was that they had merely "been 'squatters' on their own island" (Darby 2018; Simmons 2017; Suliman 2018). As bluntly stated in Gould and Lewis's article "Green Gentrification and Disaster Capitalism in Barbuda," Hurricane Irma "opened the door for the government and foreign investors to remake Barbuda as an ostensibly 'green' tourist destination for the global elite, dispossessing its residents—descendants of plantation slavery—from their collective reparations." If Barbudans are once more "relegated to serving the elites on what was formerly their land," the predictable outcome will be "greater inequality and a power shift from vulnerable locals to globally-mobile elites" (Gould and Lewis 2018:149, 153). At least up until now, Barbudans' shared ownership of the island—their only compensation for their long exploitation under the yoke of slavery—assured them a strong voice in deciding how much and what kinds of development and foreign investment to permit on their island. As of this writing, the future of the Barbudan people and their unique way of life remains unclear and very much under threat.

5

Food Provisioning at Betty's Hope Plantation, 1780s–1850s

GENEVIÈVE GODBOUT

During the eighteenth and nineteenth centuries, the residents of the Betty's Hope Plantation obtained food and various edible goods from a complex network of provisioning and trade spanning the Atlantic World, from the neighboring island of Barbuda and the garden plots cultivated on the estate's grounds to the English metropole, occasionally venturing beyond the boundaries of the British Empire despite colonial policies. Although the historiography usually depicts the eating and drinking habits of Euro-Creole plantation residents and owners in terms of an abundance and excess enabled by transatlantic trade networks, food supply on Antiguan plantations such as Betty's Hope was nevertheless fragile and vulnerable to scarcity (Bickham 2008; Petley 2012; Walvin 1997).

This chapter traces the networks through which the residents of Betty's Hope obtained food during the colonial period and the relationship between food provisioning and the social order of the plantation in the decades surrounding the abolition of the slave trade (1807) and subsequent emancipation of enslaved laborers (1834), which, in the context of this chapter, spans the 1780s to the 1850s. At Betty's Hope and other sugar plantations of the Leeward Islands, edible goods were produced and circulated at multiple scales—local, regional, and the global Atlantic—which residents could access according to their position within colonial society. Food production, trade in edible goods, and the convivial consumption of food and drink appear prominently in the archaeological and documentary records of life at Betty's Hope, and they can help us follow the thread of social transformation brought about by military conflicts, the emancipation of labor, changing relations of reciprocity between colony and metropole under absentee ownership, and the politics of garden agriculture in the Leeward Islands.

Documenting Food Supply at Betty's Hope under Absentee Ownership

The archaeological research upon which this chapter is based was carried out at Betty's Hope between 2011 and 2013 in the kitchen yard and service building complex associated with the great house. Following a comprehensive survey of the kitchen yard, two areas—approximating 16m^2 in total—were excavated, focusing on a detached kitchen (F1000) and a building identified as a "bookkeeper's dwelling" and stores in a 1755 survey (F1003; figure 5.1).

Both excavated areas yielded primary deposits dating from the late seventeenth century to early twentieth, associated with domestic occupation and food preparation. The assemblages recovered comprised primarily ceramics, including abundant industrial coarse earthenware used in sugar production (sugar molds and molasses jars), locally made low-fired coarse earthenware, refined imported earthenware, and faunal material.

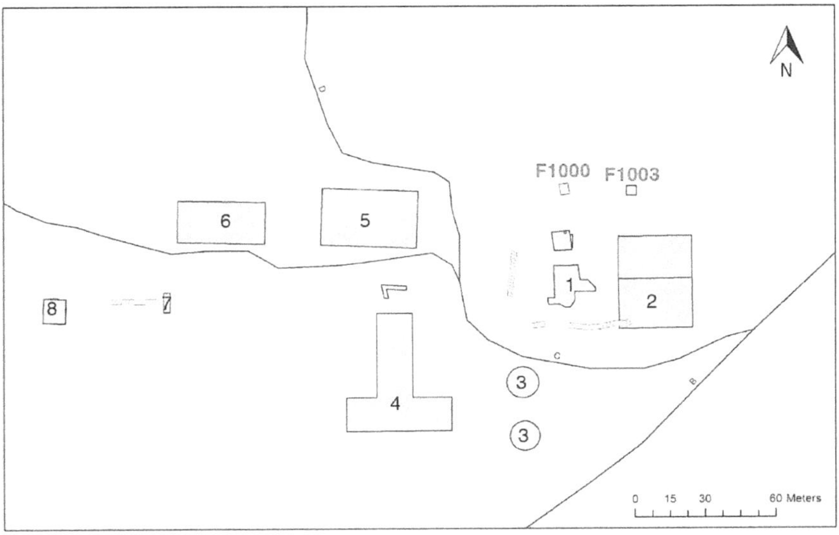

Figure 5.1. Site plan of Betty's Hope Plantation (PEH-001), showing the two shovel test areas, F1000 and F1003. Other buildings represented are as follows: (1) Great House; (2) cisterns; (3) windmills; (4) boiling and curing house; (5) Manager's House; (6) stables; (7) trough; (8) cistern. By Cory Look.

Furthermore, tableware that included valuable plates and imported ceramics was typically kept in the great house, often under lock and key, and the artifacts found in the kitchen yard tend to be associated with food preparation and storage (Arcangeli 2015).

More than 6,080 artifacts were recovered over the course of excavation and survey, in addition to extensive zooarchaeological assemblages, comprised mostly of small to large mammals and a variety of fish species analyzed by Alexis Ohman (2013; chapter 6 of this volume). Although soil samples from all primary contexts were analyzed for archaeobotanical remains, only a few charred cucurbit seeds were recovered, and as a result of the poor preservation of macro-remains at the site, overall preliminary archaeobotanical analysis was determined to be neither viable nor statistically significant.

The period considered here was well represented throughout the archaeological record in the Betty's Hope kitchen yard, which was occupied continuously from at least 1710 to the early decades of the twentieth century. During that period, the proprietors of Betty's Hope, successive members of the Codrington family, oversaw their estates from afar, having resettled permanently to England in 1720, generating an extensive archive of plantation accounts and correspondence between the Codrington owners and the various attorneys and managers who resided at Betty's Hope. These documents, consulted at the British Library (Codrington Correspondence surrogates, RP2616), the Gloucestershire Regional Archives (Codrington Family of Dodington, D1610), the National Archives, Kew, England, and the Museum of Antigua and Barbuda, are considered here alongside archaeological evidence to build a detailed picture of changing food provisioning practices at Betty's Hope during the colonial period.

Some key findings from the analysis of provisioning practices over several decades are synthesized below to illustrate how accessing imported edible goods and tableware was part and parcel of how plantation resident attorneys and managers asserted their relationship to the absent proprietors. This is opposite to plantation laborers, who tended to assert their position as food suppliers through local garden agriculture.

Although the island of Antigua thrived on plantation agriculture during most of the period considered, as early as the mid-seventeenth century, sugarcane had all but monopolized the productive output of the island, relegating subsistence agriculture for the most part to relatively small-scale garden plots scattered across the island, in enslaved laborers' quarters, and near great houses. Generally, the people of Antigua, regardless of their position within plantation economies and colonial bureaucracies, depended on imported goods for their subsistence, which included salted cod from Newfoundland, cattle and grain from the American colonies, and water from the neighboring islands of Montserrat, St. Kitts, and Guadeloupe.

A survey of the plantation accounts from the eighteenth and nineteenth centuries shows that Betty's Hope received a wide variety of foods purchased from wholesalers in London, Bristol, or Liverpool. Foodstuffs included staples of a European diet such as flour, peas, and butter. Butter was imported from England in wooden firkins, which did little to ensure the product would arrive unspoiled after weeks of humidity and heat. The rancid taste of butter in the West Indies was one of the first things Lady Georgiana Codrington noted about Caribbean food upon her arrival in the region in 1843. She wrote that "every dish was well soaked in butter (the Antigua butter being quite poisonously nasty) it may easily be conceived that the smell or taste of the cookery was not enjoyable" (Codrington 1843–1844:54). Despite its purported toxicity, butter remained part of the diet at the Betty's Hope Great House and in Antiguan Euro-Creole kitchens more generally, notably as a sauce for "dressing" fresh sea turtle meat (Codrington 1843–1844:54; Lanaghan 1991 [1844]:137).

The residents of the Betty's Hope Great House also consumed another staple of English foodways, beer and ale, albeit on a comparatively modest scale, particularly in a context where rum flowed abundantly (see also Goudge in chapter 12 of this volume; Smith 2005). Beer and ale were imported to the Betty's Hope estate directly by the Codrington proprietors, who sent an average of six to eight dozen bottles of beer every year throughout the 1740s ([BL]:Codrington Papers [CP] RP2616:8; Gloucestershire Archives [GA]:Codrington Papers [CP] D1610:A2) and then more sporadically until at least 1810 (BL:CP RP2616:3; GA:CP D1610:A8). Beer was shipped to Betty's Hope either in wooden casks or, for smaller quantities, in stoneware bottles and growlers, the vessels in which beer was commonly available throughout the Anglo-American world, although it is unclear whether wooden casks or stoneware bottles were preferred at Betty's Hope.

Wooden containers are rarely preserved in the archaeological record, and as such, beer and ale bottle fragments are among some of the most

Figure 5.2. Stoneware bottle fragment from BH2012-STU100-6-22, shown with associated assemblage.

common evidence for alcoholic beverages in archaeological assemblages of the British Atlantic, along with bottle glass. At Betty's Hope, less than a dozen stoneware beer bottle fragments were found among the hundreds of artifacts recovered from the shovel tests conducted in the area of the kitchen yard, excavated between 2011 and 2013 (figure 5.2). Although the quasi-absence of beer bottles in the Betty's Hope assemblages suggests that wooden casks might, after all, have been the preferred container for beer importation at Betty's Hope, it also indicates that the consumption of beer and ale might have been reserved for significant social occasions, particularly considering that other drink containers, notably glass bottles, were far more common.

In addition to beer and ale, coffee, chocolate, tea, and wine were imported to Antigua on account of the Codrington proprietors throughout the eighteenth and nineteenth centuries, making their way to Betty's Hope in various quantities and at various frequencies. The ships bringing plantation stores to Antigua also stopped at Madeira to pick up wine (e.g., Richard Oliver to Sir William Codrington, April 1783; BLCP Correspondence:31; Codrington Correspondence:C14/1, p. 25), which was for the most part meant to be sent back to the Codrington family with the shipment of sugar, although some Madeira wine was consumed in the context of convivial hospitality at Betty's Hope. Libations might have also included beer, gin, and other imported liquors, as well as rum, the ubiquitous Caribbean spirit often consumed as punch, a mixture of fruit juice, rum, aromatics, and sugar served in large ceramic bowls. Both punch and punch bowls were emblematic of West Indian sociality, which prompted punch bowls to appear commonly in political cartoons on the Caribbean colonies and in the bar equipage of inns and coffee houses throughout the British Empire (Harvey 2012).

Salted and barreled meat products, including beef and pork, were also purchased from various English retailers, particularly in London, Bristol, and Liverpool, or in some cases from Ireland, and shipped from the metropole to Betty's Hope. This was to supplement the rations given to enslaved laborers, particularly at Christmas, but also to feed the resident managerial staff year-round. In addition, the resident attorney-manager at Betty's Hope received a variety of stock and produce from the neighboring island of Barbuda, a Codrington dependency from 1684 to 1876 (as in chapter 4 of this volume; Dyde 2000:194; Sluyter 2009). The provisions sent from Barbuda to Betty's Hope during the eighteenth and nineteenth centuries included tropical fish and various types of produce, including Guinea corn and yams, as well as various poultry and meat that, according to Ohman's analysis of the faunal assemblages recovered from the Betty's Hope kitchen yard, was most likely butchered off site.

Tuber, maize, and Guinea corn were for the most part distributed among the enslaved residents of the plantation as part of their rations, but archival evidence suggests that most of the meat sent from Barbuda to Betty's Hope was destined for the great house kitchen. In 1783, the former resident attorney-manager of Betty's Hope, Richard Oliver, instructed his replacement, Joseph Lyons Walrond, that throughout his residence at Betty's Hope, he would be entitled the following provisions from Barbuda:

> 24 fat sheep, 12 lambs, 12 kids, 6 [female] goats, 2 doz fowls to be sent at your expence [*sic*] to Barbuda to breed and fed there beyond the 2 doz now there. 12 turkeys also to be sent by you for breed & fed for you [12] small turtle or otherwise in weight equal to 360 pd an necessary wood for the kitchen at BH, frog according to luck & when opportunities offer of sending it to you, much as crabs, lobsters, wild fowl, venison or other game—the whole of the

> above an annual supply according to your desire when opportunities offer to the [moment]. (BLCP Correspondence:31; Codrington Correspondence:C14/1, p. 1)

In addition, resident attorneys and managers employed by the Codrington family in Antigua and Barbuda benefited from the salvage and sale at auction of the cargo from ships wrecked off the coast of Barbuda, which often included edible goods such as gin, pickles, and flour, which might have been in short supply on Antigua, as well as a variety of supplies that were destined for other markets and, as such, might not have been otherwise available in Antigua. For example, in March 1784, the contents of the ship *De Neutralité* included shipments of fine silk destined to New York, which were auctioned off to an eager crowd (BLCP Correspondence:19; Codrington Correspondence:E20).

Changing Providers at the Betty's Hope Table

Between the 1780s and the 1840s, food supply networks at Betty's Hope underwent several transformations, some relating directly to absentee ownership, notably the decline in provisions sent from Britain to the resident attorney-managers in the great house. In 1783, as Joseph Lyons Walrond took his residence at Betty's Hope, he fully expected that the Betty's Hope Great House would be furnished by the Codrington proprietors (including tableware, cooking utensils, and furniture), and that the latter would send various provisions such as flour, grains, pickles, and medicine, selected and purchased for the most part in London by the Codrington family agent in the city. By 1848, however, a short generation after Emancipation and well into the transformation of British and colonial Atlantic economies in the wake of the Industrial Revolution, the resident attorney-managers at Betty's Hope no longer received food or tableware from the Codringtons, although the latter still purchased linseed meal and other products to feed and care for the cattle and working animals of the plantation.

This change is paralleled in the archaeological record of the Betty's Hope kitchen yard by an uptick in the relative proportion of Scottish refined earthenware. For example, in a shovel test unit where artifacts dated to ca. 1840s–1860s, fragments of polychrome sponge-decorated refined earthenware likely associated with Scottish manufacturers accounted for approximately 64% of the total ceramic assemblage, which would be consistent with a turn to local, predominantly Scottish retailers in Antigua for household supplies. Indeed, many local retailers of Antigua set up shop in St. John's on a street nicknamed "Scotch Row," denoting the metropolitan origin of the merchants and from whom the managers residing at Betty's Hope certainly obtained provisions and other goods (Dyde 2000:165; Lowes 1995:12). Scotch Row merchants also sold a variety of provisions and dry goods, including flour, grains, tea, and coffee, that the Codrington proprietors had sent at Betty's Hope through the eighteenth century and that the occupants of the Betty's Hope Great House were likely to obtain through Antiguan resellers from the 1840s onward.

Nevertheless, far from having relied exclusively on Codrington imports to supply its inhabitants with food and cooking implements, the Betty's Hope Plantation had a long history of drawing from a variety of local supply networks, particularly fresh produce, but also cooking implements and tableware. The staff of the great house grew vegetables and tubers in gardens adjoining the house, raised poultry, and tended to tropical fruit trees such as tamarind and lime located on the estate to feed the resident attorney-manager, his household, and his occasional guests. They could also purchase other produce and goods at Sunday markets, continuing to rely heavily on the "internal market" system already well established in Antigua since the eighteenth century (Tomich 2000) and which depended on the labor first of enslaved Antiguans and, after 1834, of their descendants and of new migrants.

This internal market in edible goods (produce, legumes, fish, small stock, herbs, and sweets) centered on the Sunday public market, particularly at Otto's Pasture in St. John's and in smaller venues such as Parham or English Harbour, where workers and agriculturalists, predominantly African Antiguans, offered produce and small stock for sale, raised on estate provision grounds and personal garden plots. These gardens relied on intercropping, or "compound gardening," in which ground provisions (yams, ground nuts), vegetables (gourds, greens, eddoes, beans), legumes, fruits (bananas, plantains), herbs, and other crops grew alongside each other (Twitty 2011:245). Under slavery, Antiguan workers began supplying the island's internal market with produce they had grown or raised in their gardens, through which they gained access to a terrain of autonomous economic activity outside of plantation agriculture, which they maintained well after Emancipation and into the twentieth century (Heath and Bennett 2000; Thomasson 1994).

From the point of view of plantation management, garden agriculture and the involvement of enslaved Antiguans in food production within the island also seemingly relieved the logistical problems stemming from shortages and rationing on the plantation. As such, unfree Antiguans had customarily been allowed garden plots since the early days of slavery, although this practice was written into law only with the 1798 Act for Ameliorating the Situation of the Slaves, which purported to improve the conditions of slavery on the island, formalize the distribution of food rations among enslaved laborers, and formalize the need for "provision grounds" (Legislature of Antigua 1805; see also Dyde 2000; Gaspar 1985; Lightfoot 2015; Lowe 1951).

The system of garden agriculture that emerged from plantation provision grounds over the centuries established a resilient system of garden agriculture that provided the enslaved population with a solid foothold in internal marketing and parallel economies, as well as a venue in which to build community ties across plantation estates (Beckles 1989; Hauser 2008; Mintz and Hall 1960; Pulsipher 1990, 1994; Tomich 2000). Contemporary descriptions of market scenes, as well as plantation accounts, suggest that Sunday market stalls were predominantly held by women, who were otherwise denied access to most skilled craftsmanship positions, which remained reserved for men who labored as carpenters, coopers, sailors, blacksmiths, and strikers (who determined when sugarcane sap had been sufficiently boiled down and was ready to be potted). Both modes of employment, as market vendor and skilled laborer, were built into the regimens of labor maintained on Antiguan plantations, in which the food grown and raised by enslaved Antiguans and their descendants was a vital source of fresh produce for Afro-Antiguan and Euro-Creole populations alike, and that also provided a means for enslaved Antiguans to enter the cash economy (Beckles 1989:45–47, 79; Hauser 2008; Pulsipher 1990). Internal marketing based in both currency and bartering helped maintain a sort of parallel geography that often exceeded the confines of plantation labor and the restricted patterns of movement and socialization it imposed (Lightfoot 2015).

As such, throughout the Caribbean, garden agriculture is associated with forms of cultural resistance and resilience based in agricultural practice and the commercial exchanges it made possible. Garden agriculture also highlights the contribution of enslaved persons to the adoption of exotic tropical species in the Caribbean, such as eddoes and groundnuts, intertwined with the perpetuation of botanical knowledge carried over during the Middle Passage (Carney and Rosomoff 2009; Kelly and Wallman 2014; Sluyter 2012).

Sunday markets and the ability to sell produce openly on that day was one of the most fiercely guarded customary rights that unfree Antiguans had carved out of the brutal exploitation of slavery. In March 1831, on the eve of Emancipation, Antigua's legislators attempted to ban Sunday markets "in observance of the Lord's day," which would have effectively prevented Antiguan laborers from doing business on their only day off and therefore from continuing to draw revenue from their surplus agricultural production, as they had done for decades hitherto. Antiguan vendors, laborers, and hucksters protested the proposed legislation by rioting and setting fire to cane fields and bagasse heaps across the island. Although the Sunday market ban was never formally lifted, by 1832, market day had quietly shifted to Saturday, where it would remain until the close of the nineteenth century (Lightfoot 2015:67–68, 80; on arson, see also Richardson 2004).

Gardening continued after Emancipation, as did the practice by plantation management of allotting garden plots to resident laborers who continued to occupy the same housing structures and settlements first established under slavery (Holt 1992; Lightfoot 2015; Smith and Smith 1986). Antiguan legislators further leveraged the dependence of the newly freed laborer on garden plots for self-provisioning, as a means for keeping the labor force in place after Emancipation. As such, even after Emancipation, because of the scarcity of free agricultural land on the island, Antiguan workers were often faced with the choice of remaining on the plantation to which they had been bound to retain both their access to housing and to a garden or seeking employment elsewhere and jeopardizing access to both shelter and independent food supply.

Atlantic Neighbors: Trading with the American Colonies and Guadeloupe

Throughout the colonial period, the occupants of the Betty's Hope Great House also looked to their Atlantic neighbors to establish a valued provisioning lifeline, particularly before the advent of steam transportation and often in defiance of the Navigation Acts that prohibited trade between Britain or her colonies and the enemies of the British Crown, including their colonies. Because British goods had to be shipped across the Atlantic in often perilous voyages, American products tended to be more reliably obtained in Antigua, as they were less likely to be lost to storms, piracy, spoilage, or damage. As such, the experience of the American Revolution and its repercussion on trade and provisioning varied throughout the Caribbean and within the Leeward Island group as well. In the case of Antigua, a potential exclusion of American colonies by the metropole from British colonial networks of trade carried the threat of a subsistence crisis.

Following the Seven Years' War, trade between Antigua and the American Eastern Seaboard, which had flourished during the first half of the eighteenth century, had first been hindered by various new colonial taxes including the Sugar Act (1764) and the Stamp Act (1765, further extended in 1775). North American traders also boycotted the British islands that complied with the Stamp Act and refused to send colonists any outbound cargo (Bemis 1962:32–33). These restrictions would transform the long-lived and reciprocal trade partnership enjoyed between the two regions by pushing them to pursue widespread illicit practices for over four decades. The estimated produce and stores imported from America into Antigua between 1787 and 1792 averaged 15,192 vessel tons a year (BLCP Correspondence:32; Codrington Correspondence:E25). At Betty's Hope, between 1787 and 1792, and indeed throughout the eighteenth century, some of the most traded items besides Guinea corn included peas, rice, oats, lumber, staves, shingles, metal hoops (to make the hogsheads in which sugar was shipped), and some livestock.

In 1783, after American independence, the British Parliament swiftly closed off direct trade between its Caribbean colonies and the United States within the framework of the Navigation Acts, first established in the seventeenth century and renewed through 1849. British Leeward Islands colonies, such as Antigua, feared that the interruption of American supplies, already diminished by the war, might provoke severe famines that, in turn, would nurture unrest among the slave population. The enslaved population was indeed the first to see its rations diminished and to suffer from hunger in times of scarcity; malnutrition, it may be remembered, was one of the leading causes of the exceptionally high slave mortality in

Antigua in 1780 (Brown 2008:56; Dyde 2000:96; Lanaghan 1991 [1844]:114; Sheridan 1974). As such, despite the risks of retaliation, many of the Leeward Caribbean islands continued to trade with the United States after 1783, and indeed well into the nineteenth century (Carrington 2000:343; O'Shaughnessy 2015).

Overall, the last quarter of the eighteenth century saw a steep inflation in the price of food, livestock, and lumber across the Caribbean, and the rising costs of operating a sugar plantation in Antigua are readily visible in the Betty's Hope accounts (BLCP Correspondence:23; Codrington Correspondence). For example, between 1775 and 1782, the resident attorney-managers of Betty's Hope, Samuel Redhead (1759–1779), Richard Clarke (1779–1782), and Richard Oliver (1781–1783), reported an overall increase of roughly £3,695 (local currency) in expenditures in procuring food rations for slaves—including barreled pork, herring, corn, and peas—due to "the very high price of articles formerly had from North America" (BLCP Correspondence:23; Codrington Correspondence:A23, Plantation Accounts, 1782, p. 61). Considering that the average operating cost of Betty's Hope in the 1760s was approximately £20,000, an increase of some £3,700 would have been significant. In the 1782 letter, manager Richard Oliver also hints at a "risky" trade he maintained with New York in order to dispose of the estate's sugars through the "usual" conduits—and also quite likely to obtain stores and provisions through unregulated trade routes.

In colonial Antigua, illicit trade was not only a palliative to the food insecurity that characterized the last quarter of the eighteenth century, it was also part of a longer trend in practices of provisioning, arguably best exemplified by the trade in freshwater between Antigua and its neighbor to the south, the French island of Guadeloupe, particularly in times of prolonged drought like the one that plagued Antigua throughout the 1770s and 1780s. During the colonial period, water supplies on Antigua depended chiefly on water cisterns, and considering that water shortages were usually accompanied by bouts of dysentery and other illnesses, supplies of potable water were invaluable (Berland et al. 2013). As such, when cisterns began to run low, it was common for Antiguan planters and merchants to obtain freshwater from the neighboring islands of Montserrat, St. Kitts, and Guadeloupe. Although technically a violation of the Navigation Acts, this practice is documented throughout the eighteenth century, likely having begun as early as the late seventeenth century (Dyde 2000).

Figure 5.3. French faïence from PEH-001-BH2013-STU106-11.

In 1789, Luffman writes in his letters that claret is also "smuggled from our French and Dutch neighbours at Guadeloupe and St. Eustatia" (Luffman 1789:51).

At Betty's Hope, hints of these expeditions to Guadeloupe are found in fragments of French Delftware, with a light reddish-pink fabric and milky enamel with minimal decor consistent with the ceramic production of northwest France, recovered from contexts of the kitchen yard associated with the mid-eighteenth to early nineteenth centuries (STU106-4 and STU106-15) (figure 5.3). This find is unusual not only because approximately 90% of the Delftware found during the excavation of the Betty's Hope kitchen yard was of English or Dutch manufacture, but also because the Navigation Acts, through most of Antigua's colonial history, would have prevented French-made ceramics from entering most British colonial markets to a significant degree. It is possible, however, that a limited quantity of such ceramics entered circulation in Antigua and at Betty's Hope via the neighboring island of Guadeloupe, where the ongoing trade in French water was likely accompanied by various other products (and their containers). In the British Atlantic, Delftware was most popular for tableware (such as punch bowls, plates, and jugs) and to a lesser extent for storage vessels and ointment jars (Genet 1996; Jelks 2007; Métreau 2017; Waselkov and Walthall 2002).

Concluding Thoughts

In sum, at Betty's Hope and other plantations of the Leeward Islands, what people ate and how they obtained it depended in great part on their position within the political and productive order of plantation slavery and its legacies in the British Empire. The residents of Betty's Hope accessed various provisioning outlets—including their own gardens, neighboring Atlantic colonies, private imports from the metropole, and local retail in public markets—distributed along the far-reaching axis of trade that moored Antigua to Britain and the American continent—within the context of the Atlantic World. Although the examples presented here were drawn from a close examination of the service area surrounding the plantation's dwelling house, archival and archaeological evidence both show how the residents of Betty's Hope had access to varied and complementary provisioning networks, including slave gardens, imported foodstuffs, and the Barbuda provision grounds. In addition to its nutritional and gustative properties, food is indeed open to culturally and historically specific grammars of edibility, encoded by colonial geographies and ideologies of consumption and made possible by economies and regimes of labor that support food production and distribution (Douglas 1972; Goody 1982; Mennell 1996), As such, the various actors who participated in plantation economies enjoyed different kinds of access to the networks of trade that brought edibles to and from Antigua. They also imagined themselves to be the legitimate recipients of different goods, in relation to the influence they could exert over the quantity and quality of provisions directed to them and which they sent outward.

Another running thread in the history of provisioning at Betty's Hope is the often-disruptive yet transformative impact of military conflict and shifting colonial policy, particularly in regard to the emancipation of labor, the continuing importance of subsistence agriculture to weather the interruptions of global trade, and to the changing relationships between colony and metropole. In the case of Antigua, the place of North American colonies as a third party to this relationship in the late eighteenth to the mid-nineteenth centuries remains overall underexamined, both under slavery during the American Revolution and after Emancipation in Antigua during the American Civil War. A closer examination of trade relationships between the Betty's Hope Plantation and the American continent, for example through the lens of stoneware containers and the edible goods they contained, could not only yield a finer-grained analysis of the Betty's Hope kitchen yard assemblages, but also shed light on the personal and familial bonds that united merchants, traders, and planters across the imperial divide.

6

Not Much Ado about Mollusks

Zooarchaeological Analysis of Mollusks at Betty's Hope Plantation in Antigua, West Indies

ALEXIS OHMAN

The analytical potential of mollusks is well known for pre-Contact and prehistoric faunal assemblages. From the clam gardens of British Columbia to the conch middens of the Caribbean, mollusks provided a foundation for human subsistence throughout time and across cultures (Claassen 1998; Deur et al. 2015; Keegan and Carlson 2008:61–70; Newsom and Wing 2004:24–25). Yet in historical zooarchaeology, the focus has often been on the myriad roles domesticated animals played in bridging Old World and New World foodways practices (for example, Bowen 1975, 1998; Crader 1984, 1990; Klippel 2001; Landon 2009:85–89). These studies have been crucial to the development of historical zooarchaeology as a subdiscipline and to historical archaeology more broadly. Yet the realm of the wild was often used as a point of contrast rather than the central focus of analysis. Resources procured from these wild spaces are often treated as lump units, listing pertinent environmental and ecological information for the taxa but often lacking further discussions (exceptions include Bowen 2000; Dawdy 2010; Scott 2001, 2008; Sichler 2003; Wallman and Grouard 2017).

Mollusks particularly suffer from this bias in historical zooarchaeology, often simply dismissed as a bulk protein source with little additional analytical value unless the shell itself is modified. However, different cultural groups will react to, integrate, or spurn certain species depending on how their foodways practices are expressed. The term foodways encompasses the full process of turning plants and animals into food, and in turn, the interrelated stages that can be influenced by sociocultural concepts and activities along the way. These stages include production/acquisition (hunting, collecting, raising), processing (butchering, cooking, preserving), consumption, and disposal of refuse (Twiss 2012:361–362).

Such processes require socioculturally informed decisions throughout the entire process of deciding what is good to eat, acquiring it, and consuming it—even for such seemingly simple creatures as mollusks. Different species will reside in different ecological niches, which requires decision making and local ecological knowledge of what is safe to eat, good to eat, and socioculturally acceptable to eat. The faunal analysis of mollusks from Betty's Hope Plantation will place a spotlight on this often-overlooked taxon, expanding discussions of historical zooarchaeology for plantation sites within the ecological constraints of small Caribbean islands such as Antigua.

The mollusk assemblages from Betty's Hope are from three spatially discrete contexts that also correlate with the plantation hierarchy, which was divided along race- and class-based lines. This provides two distinct goals. First, a focus on mollusks will push current historical zooarchaeology in the Caribbean to consider wild resources as important in their own analytical right rather than a dichotomous comparison between domestic and wild species. Second, the focus on mollusks will permit a discussion that specifically integrates ecological knowledge with economic ability and the social mores of what is considered good to eat.

Antigua and Barbuda provide an excellent backdrop with which to examine tropical marine resource use. The small islands in the Lesser Antilles were exceptionally susceptible to the ecological changes wrought first by the indigenous communities, but more dramatically by the European colonists, the plants and animals they brought with them, and the bourgeoning sugar-industrial complex. The importation of new animal taxa from Europe, Africa, and Asia such as cattle, sheep, goats, pigs, chickens, guinea fowl, rats, and mongooses caused extinctions and extirpations (local extinctions) of terrestrial animal taxa. In the wake of these severe limitations of terrestrial resources, the remaining local taxa that could be relied upon were fish and mollusks to supplement plantation rations (Dyde 2000:4–5; Harris 1965:19–23).

In conjunction with these ecological conditions were significant cultural habits of the English that affected how they interacted with the tropics and managed their provisioning systems. They preferred familiar, Old World domesticates such as sheep, goats, pigs, cattle, and fowl as the foundation of their diet. Luxurious exceptions such as sea turtle did become incorporated into fine dining on English tables (Dunn 1972:276). By the seventeenth century, fish had fallen into particular disfavor, though the remaining preferred species tended to be fattier, oilier, red-fleshed fish that more closely resembled red meats. The tropical Caribbean environment compounded this pattern of consumption because the English also had a particular fear of hot climates, therefore resulting in a distaste for many tropical resources that were considered inferior at best and dangerous at worst (Dunn 1972:263–264; Franklin 1997; Jarvis 1997; Kupperman 1984; Newsom and Wing 2004:215; Scarry and Reitz 1990).

This cultural rigidity was of course not absolute. Research conducted since the 1970s has demonstrated some of the various ways British, French, and Spanish colonists included wild taxa into their diets (e.g., Bowen 1975, 2000, 2006; Cleland 1970; Reitz 1979; Reitz and Honerkamp 1983; Scott 2001, 2008). French colonial sites often modify their foodways practices to include local wildlife, which was consistent with their homeland tendency to include more wild game in their diets compared to their British and Spanish counterparts. This practice persisted in the Caribbean, including acquiring indigenous knowledge of capture methods for these new resources (Price 1966; Wallman and Grouard 2017).

Yet these studies are difficult to apply to islands such as Antigua, which have such a small landmass that potential animal populations are substantially restricted. The abundant wild game and indigenous trade networks other colonists utilized on mainland North America were simply not available on small Caribbean islands (Newsom and Wing 2004:1–3). Fish and mollusks were the remaining local taxa to reliably exploit, but procuring these creatures was further hampered by the rapid spoilage of meat in the tropical heat. This was such a problem that Caribbean markets would forbid selling fish after the morning hours (Dunn 1972:275–276). It was therefore more practical for planters to rely on the long- and short-distance provisioning systems rather than allocate labor for marine resource collection, which detracted from the labor requirements of sugar plantations such as Betty's Hope.

The shortage of usable land in the Lesser Antilles, coupled with the English disdain for local tropical resources in favor of the familiar Old World domesticates, resulted in both a lack of understanding of local ecologies and a dependence on long- and short-distance provisioning systems. Tropical fish and mollusks were left to those who were less able to afford or have access to the desirable imported goods. Marine subsistence activities by the non-elite on the plantations would not have been reported in the Codrington archives because they rarely affected the Codrington profit margins—though some exceptions will be noted in the following section. Historical zooarchaeology is therefore perfectly poised to provide a bottom-up approach to understanding these invisibilities in the archival record. The restricted niches that both fish and mollusks occupy can also inform on where these animals were procured from beyond the boundaries of the provisioning system.

Animals in the Archives

Research at Betty's Hope Plantation has benefited substantially from the comprehensive historical archives produced by the Codrington family and the plantation managers and attorneys who supervised Betty's Hope and the other Codrington estates on Antigua and Barbuda. This collection of letters, accounts, and shipments is collectively called the Codrington Papers. The following section will refer to these archives to discuss the various taxa utilized at Betty's Hope prior to delving into the zooarchaeological analysis of mollusks.

Domestic mammals were at the forefront of the Codrington Papers because of their significance to daily life on Betty's Hope Plantation and the family's other Antiguan estates. Consequently, livestock was also central to Barbuda; the majority of the domestic animals were bred and maintained on this island rather than on the Codrington estates themselves. Some were actively managed, but the majority were left to become feral on Barbuda until they were needed (chapter 4, this volume; Dyde 2000:38; Reitz 1992:7; Watters 1997:231). Fowl occupied a couple of different niches. Domestic birds such as chickens and guinea fowl were managed in flocks, which also likely included ducks, geese, and turkey. Other birds were stocked on Barbuda for sport shooting, including such exotics as flamingos and the West Indian whistling duck (NAAB Codrington Papers:C16; Reitz 1992:7). The degree to which these birds were also desired for their meat is not clear.

For marine resource use, the archives are quite sparse. As previously indicated, sea turtle was highly valued on both economic and sociocultural levels. Turtlers captured droves of green and hawksbill turtles to sell the meat, eggs, and shell. Live animals were also shipped abroad to become turtle soup in the tureens of elite English households, but these creatures often died along the way (Dunn 1972:276; NAAB Codrington Papers:C16; Wilson 1973:203–204). When this happened, "mock-turtle soup" used seasoned calf's head to replace the deceased turtle but retain the air of prestige (Wilson 1973:204). Yet even these discussions do not get at the question of how non-elites interacted with other tropical resources on a daily basis. The only fish species explicitly identified in the archives were imported salted herring and cod, and even a quantity of salmon salvaged from a wreck off Barbuda is mentioned. Although local fish were not identified by species, a handful of notes in the Codrington Papers do discuss the development of local fishing practices on Barbuda that demonstrate a shifting understanding of available marine resources.

The best example is an exchange between Christopher Bethell Codrington and the manager of Barbuda at the time. In 1809—well over 100 years since the Codringtons took over Betty's Hope—the manager did not believe that there were enough fish in the waters around Barbuda to supply the hundreds of slaves owned by the Codringtons (NAAB Codrington Papers:C24). This is also evidenced by extensive calculations in the Codrington accounts rationing barrels of salted herring for the enslaved Africans at Betty's Hope and the other Antiguan estates (NAAB Codrington Papers:C19). By 1824, it was noted that fishing had become commonplace on Barbuda—primarily to supply the enslaved Africans there—and that barrels of surplus salted fish were shipped to Antigua (NAAB Codrington Papers:C24; Ohman 2014:62–63).

The few instances where mollusks are mentioned refer to the shells rather than the meat, and only two of those references were for a specific species. The first involves a 1720 request for "some Conkshells," which likely refers to the ubiquitous queen conch (*Lobatus gigas*) (NAAB Codrington Papers:C1–6). The second is a letter dated between 1789 and 1791 that requests "a Box of Mangrove Oysters with pieces of the Mangrove trees they grow on." It is explained that these mangrove oysters (*Crassostrea rhizophorae*) and "a Box of Stone Specimens as p list inclosed" are desired because they are "curiosities" (NAAB Codrington Papers:C29; Ohman 2014:49–58).

A third mollusk should be noted for both its importance in Caribbean foodways and its ornamental appeal, even though it was not mentioned in the archives. The West Indian top shell (*Cittarium pica*) simply put, is a beautiful shell. The exterior of this snail's shell is mottled white and deep purple, with a pearlescent white interior. It can reach 10–12 cm in diameter, and today they are commonly featured along with conch as decoration in homes, restaurants, and outside as ornaments. The snail itself can be consumed either hot in a soup or stew or sliced cold in a salad-style dish (Abbott 1974:48; Robertson 2003:40–41). It is therefore quite likely that such a prominent and easily recognizable shell would also be requested for souvenirs.

Despite these rare glimpses into mollusk use in the Caribbean—be it on the plate or on the shelf—the faunal assemblages recovered from Betty's Hope demonstrate a much wider variety of mollusks than would be expected based on the archives alone. In all likelihood, this can be attributed to the unavoidable nature of the archives themselves; on both sides of the Atlantic, the Codrington family and their representatives were part of the upper echelons of society, and the plantations' operations and profits were thus at the heart of the majority of the correspondence.

At Betty's Hope, the mollusks are a key way to demonstrate a stark absence in the archives that zooarchaeological analysis can explore. Mollusks are more than simply the base of a stable protein source; different species are procured from a variety of ecological niches that require both knowledge of the local environment and specific planning to procure taxa from those specific areas. The taxa also have to be deemed acceptable to eat, and they may be incorporated on different scales of value depending on class and socioeconomic status. Selecting mollusks, as with any other animal resource, is integrated not just in sociocultural choice but also within the broader environment beyond the boundaries of the plantation and into the wild spaces.

Mollusks at Betty's Hope

The results presented in this chapter are part of an ongoing analysis of the faunal assemblages from Betty's Hope that began with Ohman (2014). All mollusk specimens were identified to the lowest possible taxonomic level. The primary data gathered for each specimen were number of identified specimens present (NISP) and weight in grams (g). Secondary data such as minimum number of individuals and meat weight are still being assessed at the time of writing. For the purpose of this volume, the full species lists have been abbreviated to focus on the taxa that are most abundant and most relevant to this chapter.

Table 6.1. Quantifications for Mollusks Recovered from Three Contexts at Betty's Hope Plantation

Context	NISP	%	Weight (g)	%
Great House	403	3.0	1,249.4	5.4
Service Quarters	2,170	16.2	2,449.6	10.6
Slave Quarters	10,787	80.7	19,461.0	84.0
TOTAL	13,360	100	23,160.0	100

A total of 13,360 mollusks were identified at Betty's Hope Plantation across three different primary contexts: the great house, the service quarters, and the slave quarters. The great house contained just 3% of the total mollusk assemblage (n = 403; table 6.1). The service quarters contributed 16% of the total NISP (n = 2,170). In exceptional contrast, the slave quarters contributed 81% of the total molluscan assemblage with n = 10,787 NISP.

Great House

The great house at Betty's Hope was the primary residence for those in charge of the plantation, whether they were members of the Codrington family in the formative years of the late seventeenth century and the earlier part of the eighteenth century or the attorney-managers who supervised plantation operations thereafter (chapter 2). The majority of the great house has been excavated at the time of writing, and so the molluscan assemblage here is considered essentially complete, yet the assemblage size is paltry (n = 403; table 6.2). This demonstrates that the great house context simply did not rely on mollusks in their foodways practices, likely because they were not as esteemed as other types of foods such as red meats, as previously discussed. Because the residents of the great house would have been financially better off than other occupants of the estate, they could have afforded to consume a wider variety of more desirable goods. However, there are still some patterns worth noting in terms of the relative abundance of mollusks recovered from this part of the site, which provides comparative data for the service quarters and slave quarters.

Twenty-three mollusk taxa were identified in this context, but just seven taxa made the strongest representation, together totaling 80% of the great house assemblage. The most abundant by far was the mangrove

Table 6.2. Abbreviated Species List of Mollusks Recovered from the Great House at Betty's Hope Plantation

Taxon	Common Name	NISP	%	Weight (g)	%
Isognomon spp.	Tree oyster	22	5.5	14.1	1.2
Crassostrea rhizophorae	Mangrove oyster	154	38.2	432.8	35.7
Phacoides pectinatus	Thick lucine	83	20.6	202.3	16.7
Codakia orbicularis	Tiger lucine	11	2.7	63.6	5.2
Anomalocardia flexuosa	WI pointed venus	15	3.7	25.4	2.1
Cittarium pica	WI top shell	21	5.2	209.0	17.2
Lobatus gigas	Queen conch	16	4.0	265.0	21.9
TOTAL		322	79.9	1,212.2	100
Grand Total		403	100	1,249.4	100

Note: Taxa restricted to those with abundance greater than 10 NISP.

oyster (*Crassostrea rhizophorae*; n = 154). This taxon was almost entirely recovered from a single cache in the attached kitchen context at the back of the great house. This was also one of the two mollusks specifically mentioned in the Codrington Papers, and the abundance here suggests that perhaps it was a desired food locally as well as a curiosity abroad. The next most abundant taxon by NISP is the thick lucine (*Phacoides pectinatus*; n = 83), a common clam in Antigua found in sandy shallow waters. This is followed by the tree oyster genus (*Isognomon* spp.; n = 22), which can be found alongside the mangrove oysters, and so would likely have been collected at the same time.

Two more taxa should be noted that are not particularly prominent in the great house assemblage but which become more important in the other contexts. These are the West Indian pointed venus (*Anomalocardia flexuosa*; n = 15) and the tiger lucine (*Codakia orbicularis*; n = 11). Both taxa occupy the same sandy, shallow-water niche as the thick lucine, which was the second most abundant taxon in this context. This suggests that there were likely additional selection processes at play rather than a full resource sweep of a given niche.

Service Quarters

This context was comprised of multiple structures that had different uses over time and which have collectively been dubbed the service quarters as a shorthand for encompassing the various uses of the outbuildings

Table 6.3. Abbreviated Species List of Mollusks Recovered from the Service Quarters at Betty's Hope Plantation

Taxon	Common Name	NISP	%	Weight (g)	%
Pinctada imbricata	Atlantic pearl oyster	282	13.0	53.6	2.2
Isognomon spp.	Tree oyster	167	7.7	108.5	4.4
Crassostrea rhizophorae	Mangrove oyster	36	1.7	81.0	3.3
Phacoides pectinatus	Thick lucine	50	2.3	27.4	1.1
Codakia orbicularis	Tiger lucine	199	9.2	397.3	16.2
Anomalocardia flexuosa	WI pointed venus	644	29.7	622.2	25.4
Cittarium pica	WI top shell	149	6.9	643.1	26.3
Nerita spp.	Nerite	40	1.8	62.3	2.5
Strombidae	Conch	48	2.2	187.9	7.7
TOTAL		1,615	72.9	2,183.3	89.1
Grand Total		2,170	100	2,449.6	100

Note: Taxa restricted to those with abundance greater than 10 NISP.

surrounding the great house. Excavations concentrated on two primary features, Feature 1000 and Feature 1003. F1000 was the early detached kitchen, which was later attached to the blacksmith to share the hearth. F1003 began as a cattle pen, but a room for servants was later built on that space. The room was further modified with stone floors and a porch. Additional details regarding these excavations can be found in Godbout (2016) and chapter 5 in this volume.

Faunal material from this part of the site was markedly different from the great house, both in quantity and diversity. Thirty-one taxa were identified, totaling 2,170 NISP (table 6.3). Perhaps the starkest difference was the sheer abundance of the West Indian pointed venus (*Anomalocardia flexuosa*; n = 644) and the Atlantic pearl oyster (*Pinctada imbricata*; n = 282). The remaining taxa of note are the tiger lucine (*Codakia orbicularis*; n = 199), the tree oyster genus (*Isognomon* spp.; n = 167), and the West Indian top shell (*Cittarium pica*; n = 149).

Given the different uses of Features 1000 and 1003 over time, it was suspected that the faunal assemblage would also differ, either in total abundance or diversity. However, both contexts were similar in abundance: F1000 contained 958 NISP and F1003 contained 1,195 NISP. They were also roughly similar in terms of relative abundance of specific taxa. Two patterns stand out, however: the pearl oyster was almost entirely represented in F1003, while F1000 contained the majority of the tree oyster specimens.

This suggests that kitchen yard refuse and domestic refuse may not have been particularly disparate, and perhaps the presence of these mollusks says more about preparation and discard. When cooking mollusks for consumption, the mollusk may be shucked in the kitchen and the shell discarded with kitchen refuse. However, if the cook prepares the mollusk in the shell and then serves the dish elsewhere—say it is brought back to the domestic space to be consumed—then the shells may be more likely to be deposited with that space's trash rather than returned to the kitchen to be discarded. Lastly, the pearl oysters may not have been consumed at all but acquired with the hope of obtaining pearls.

Slave Quarters

The slave quarters excavations have added a much-needed dimension to understanding foodways practices at Betty's Hope (chapter 10), and the mollusk assemblage from this context vastly exceeds the relative abundance found in the other contexts at Betty's Hope (n = 10,787). This is generally consistent with what is often found in enslaved African contexts on plantations. Archaeological and historical records from the United States reveal a wide variety of aquatic and terrestrial resources that were utilized, such as squirrels, raccoons, opossums, deer, various game birds and waterfowl, turtles/tortoises, and fish and mollusks from various fresh, marine, and brackish niches (for example, see Crader 1984, 1990; Dawdy 2010; Scarry and Reitz 1990; Scott 2001, 2008; Swanson 2009; Thomas 1998; Tuma 2006).

Discussions surrounding the use of these resources address not just the significance of the taxa themselves, but also the social importance of hunting and fishing. They are more than just risk management strategies: they are both a social bonding activity and an active subversion of the plantation power structure (Ball 1854:251–271; Thomas 1998; Tuma 2006). Although hunting, trapping, and fishing are typically discussed in these accounts, gathering mollusks could also have filled this role. The incredible abundance and diversity of molluscan material recovered from this part of Betty's Hope reinforces the importance of these overlooked taxa. Moreover, the variety of niches these mollusks were procured from again demonstrates additional decision making, rather than simply collecting everything available.

Table 6.4. Abbreviated Species List of Mollusks Recovered from the Slave Quarters at Betty's Hope Plantation

Taxon	Common Name	NISP	%	Weight (g)	%
Isognomon spp.	Tree oyster	591	5.5	442.9	2.3
Crassostrea rhizophorae	Mangrove oyster	185	1.7	297.2	1.5
Phacoides pectinatus	Thick lucine	986	9.1	1,398.3	7.2
Codakia orbicularis	Tiger lucine	2,383	22.1	3,718.9	19.1
Anomalocardia flexuosa	WI pointed venus	1,142	10.6	1,193.1	6.1
Cittarium pica	WI top shell	605	5.6	2,544.4	13.1
Nerita spp.	Nerite	201	1.9	200.1	1.0
Strombidae	Conch	533	4.9	7,092.1	36.4
TOTAL		6,626	61.4	16,887.0	86.8
Grand Total		10,787	100	19,458.7	100

Note: Taxa restricted to those with abundance greater than 100 NISP.

Thirty-five taxa were identified from this context. The most abundant taxon was the tiger lucine (*Codakia orbicularis*) with 2,383 NISP (table 6.4), closely followed by the West Indian pointed venus (*Anomalocardia flexuosa*; n = 1,142) and the thick lucine (*Phacoides pectinatus*; n = 986). Following these are the West Indian top shell (*Cittarium pica*; n = 605), the tree oysters (*Isognomon* spp.; n = 591), conch (Strombidae; n = 533), and nerites (*Nerita* spp.; n = 201).

Discussion

The zooarchaeological analysis of the mollusk assemblages at Betty's Hope has revealed some clear patterns within and between the great house, service quarters, and slave quarters at the plantation. The difference in basic abundance among the three contexts is striking in itself: the great house (403 NISP) comprised a fraction of the total number of mollusks compared to the slave quarters (10,787 NISP). The enslaved Africans at Betty's Hope were clearly much more reliant on local tropical resources than either the elite occupants of the great house or the middle-class residents of the service quarters complex. It is likely that the higher-class occupants had sufficient economic resources to allow them to consume the familiar foods they coveted from the Old World rather than the local tropical resources they distrusted.

With these attitudes in mind, it is therefore unsurprising to have seen such a paucity of molluscan material in the great house and such a corresponding abundance in the slave quarters. Mollusks would have been easy to collect and/or cheap to purchase in formal or informal market settings, as long as access to these settings could be achieved. Since Betty's Hope is not adjacent to, but not particularly far from, the coastline, it is possible that the enslaved Africans on the plantation could have been given permission to collect mollusks during their free time and/or would have exercised their own agency to harvest them without permission.

Mollusks such as the tiger lucine, thick lucine, West Indian pointed venus, and even the queen conch are all common to the shallow-water sands and seagrass beds of Antiguan coastlines. However, the mangrove and tree oysters are procured from mangrove environments, often sharing the same tree roots (Newsom and Wing 2004:141). Given the chronic paucity of freshwater resources on Antigua, this requires going to very specific zones on the island. Mangrove oysters are larger and more robust than tree oysters, yet tree oysters were the third most abundant taxon in the slave quarters (n = 591; 5%), while mangrove oysters ranked seventh (n = 185; 2%). In contrast, the great house assemblage was dominated by the mangrove oyster (n = 154; 38%). Three hypotheses are offered for this apparent discrepancy.

First, tree oysters may have been tastier than the larger-bodied mangrove oysters and were therefore preferentially procured rather than simply included in a full resource sweep of the ecological zone. Alternatively, they may have been specifically purchased in formal and/or informal market settings. Second, mangrove oysters may have been preferred by the occupants of the great house, so any approved ventures to the mangroves by the enslaved Africans may have come at the cost of turning over the mangrove oysters to the great house. This of course does not take into account the possible acts of subversion, where mangrove oysters may have been held back instead of turned over to the great house, as demonstrated in accounts of enslaved subversion elsewhere (e.g., the fishery scheme in Ball 1854:251–271). The third hypothesis is that the prevalence may be the result of ecological differences; the niches that were sought out may simply have had a greater abundance of tree oysters and fewer mangrove oysters, which would constitute a micro-niche in the broader mangrove ecological zone. The mangrove oysters at the great house may also have been simply purchased if they were desired for a specific event, or even for an individual.

However, a major caveat for both mollusks and other fish is the risk of spoilage; formal markets were forbidden from selling fish beyond the morning hours because of rapid spoilage in the tropical heat (Dunn 1972:275–276). It is therefore most likely that mollusks would have been consumed very shortly after acquisition. Given the sheer abundance of mollusks recovered, it appears that collecting mollusks was either a common activity for enslaved Africans at Betty's Hope (with or without permission) or was an inexpensive purchase from formal or informal markets.

The mollusk assemblage from the slave quarters has provided a substantial amount of data with significant analytical potential within the broader discussions of the plantation system. Although the Codrington family may not have made much ado about mollusks in their personal archives, it is clear that the abundant faunal material can help fill in the gaps of everyday life on the plantation. Historical zooarchaeology has become a subdiscipline in its own right over the past few decades, but its application to the colonial Caribbean situation has been limited compared to mainland North America. The Caribbean region is dually challenged by diverse ecologies among the islands, coupled with the various European powers that exercised their colonial rule in very different fashions that changed over time. The result is both a patchwork and a network, where each island can be examined within its ecological constraints while simultaneously being interconnected to other Caribbean islands and the greater Atlantic World. These discussions will be mutually beneficial to the expansion of historical zooarchaeology as a subdiscipline and for the historical archaeology of the Caribbean region.

PART III

Cane Agriculture and Its Legacy

7

Agriculture at Betty's Hope

GEORGIA L. FOX

When polled about their tea consumption, respondents in the United Kingdom claimed to drink at least two to three cups per day. This innocuous national pastime translates to 165 million cups per day—60.2 billion cups per year in the United Kingdom alone (UKTea & Infusions Association 2018). Since the publication of Sidney Mintz's *Sweetness and Power* in 1985, the relationship between the adoption of tea and the adoption of sugar was recognized, firmly establishing how certain commodities have come to dominate our lives in such ubiquitous ways that even calling attention to them seems redundant. Yet there was a time when sugar was a precious luxury item, craved by many but accessible to only a few. All that would change when the English devised their most profound and far-reaching experiment: the agro-industrialization of sugarcane in the Caribbean.

The transformation was slow but eventful. To better understand the trajectory of Caribbean sugar, the archival and archaeological evidence that emerges from exploring a plantation like Betty's Hope as part of the Caribbean experiment can be instructive. This chapter will explore agriculture at Betty's Hope, Antigua, in its transformation from fledgling colony to agricultural powerhouse and determine whether the planters were truly innovative or merely following a template to achieve their bottom line.

Beginnings

When Captain Thomas Warner arrived in Antigua in 1632, it was almost as if he were following a script neatly planned and written for him; all he had to do was fell the first tree and plant the first crop. The truth, however, proved far more complex and far less idealized. Antigua's early settlement period was fraught with challenges. The first colonists who farmed cash crops like indigo, ginger, and tobacco encountered several impediments to success: a heavily forested island, low population numbers, and an alien landscape of strange animals and plants, like the poisonous manchineel tree (*Hippomane mancinella*). With so few people, clearing land was almost impossible, especially with the lack of both good-quality axes and workers who actually knew how to use them. The whole enterprise defeated "tree-felling schemes in their entirety" (Edelson 2007:383; Watts 1987:154). Establishing farms and plantations in the Leeward Islands and Antigua required the patience of Job. Steep terrain and low cash flow did not help matters, nor did the French invasion of 1667–1668 (Dunn 1972:117–118). All that was to change, however, with the passage of the Act for the Encouraging and Promoting the Settling of this Island in 1668.

The act was nothing short of a miracle. In fact, it had the almost immediate effect of attracting new settlers, one of them being Christopher Codrington II (see chapter 2). The law was particularly friendly to wealthy planters interested in acquiring large tracts of land and the slave labor they required. In this regard, the 1668 law became the watershed event for establishing the sugar plantocracy and large-scale farming on Antigua, with Codrington II leading the way. By 1681, the island had become a "sea of cane." By 1748, close to 200 plantations were recorded for the island (figure 7.1; Lanaghan 1991 [1844]:50).

Establishing a sugar estate was a significant commitment of resources. Dalby Thomas estimated that in 1690, a 100-acre plantation on Barbados that included sugar works, 50 enslaved Africans, seven white indentured servants, and the required equipment and livestock, would cost £5,625, a sizable sum at the time (Sheridan 1974:264).

Cultivating sugarcane also demanded the requisite know-how and adequate skilled and unskilled labor. Antiguan planter Samuel Martin best summed it up by stating in his *Essay on Plantership* that the sugar plantation "ought to be considered a well constructed machine, compounded of various wheels, turning different ways, and yet all contributing to the great end proposed" (Martin 1785:18). Martin was a leader among a new breed of experts who produced guidelines on successfully running and managing a Caribbean sugar plantation. His *Essay on Plantership* was so

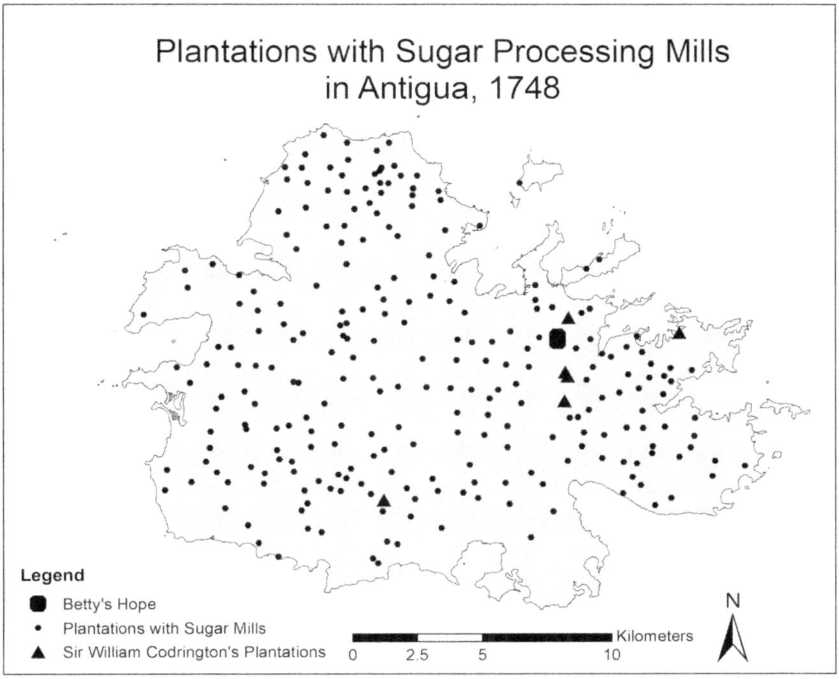

Figure 7.1. The Baker map showing the location of Antigua's plantations, 1748. Modified by Christopher Waters.

popular that the demand prompted several editions from 1750 to 1802 (Sheridan 1960:138).

Heading into the eighteenth century, developments in cane agriculture were influenced by Enlightenment values on science and technology and detailed record keeping. Planters assiduously studied Martin's manual and formed agricultural societies, with the aim to increase revenues, reduce costs, and streamline operations (Roberts 2013:31; Sheridan 1989:65). One indication of this evolving style of plantation management was the implementation of new land survey methods that shifted from the old "direct measurement" method to chain surveying, developed in 1620, and the use of triangulation and instrumentation like the box compass (Hicks 2007:43–44). The more science-based approach led to a flurry of plantation maps and plans in the eighteenth century (Hicks 2007:44). Because the Codringtons prided themselves on their forward thinking, Betty's

Hope was laid out with a certain level of preciseness, suggesting that the latest methods of land survey technology were applied.

The Plantation Layout at Betty's Hope

The eighteenth century also marks the consolidation of large tracts of land by wealthy planters; larger sugar estates like Betty's Hope measured 700 acres or more. Codrington landholdings of Betty's Hope, Cotton, Cotton New Work, and Garden estates totaled 1,426 acres (NAAB Codrington Papers:P4). When William Codrington I inherited Betty's Hope, he had the plantation surveyed, resulting in the map of 1710, which records 870 acres, much of it devoted to cane fields (NAAB Codrington Papers:P3). Between 1755 and 1782, the number see-sawed, from 543 acres, with 427 acres devoted to cane in 1755, jumping to 725 acres in 1676 (NAAB Codrington Papers:P4; Oliver 1894: 169–1899:I:125). Finally, the acreage for Betty's Hope seems to have stabilized at 543 acres from 1790 until 1927.

By the eighteenth century, Caribbean plantations were divided into discrete spaces—the domestic sphere, sugar works, water collection, animal pens, slave quarters, and support buildings—and were therefore managed like self-contained entities. Yet they were not cookie-cutter operations; each plantation was distinctive in its own right, with a layout and dimensions dictated by a number of factors like topography, available resources, and personal preference. For example, boiling houses were often placed downhill from a windmill so the cane juice could flow by gravity to the boiling house. At Betty's Hope, however, archaeological investigations of the North Mill during 1989–1990 revealed that the cane juice was collected in a cast-iron tank situated below the crushers and then pumped to the boiling house (see chapter 14).

Regardless of any differences, all Caribbean sugar estates relied on water collection. On a dry island with no natural rivers or lakes, water collection was de rigueur on Antigua. Archaeological surveys throughout the island reveal different types of water collection systems on old plantation sites in the form of ponds and various catchment systems, such as the pond at Coconut Hall. Ponds at Betty's Hope are marked on the old maps and can be seen in depressions in the plantation's landscape.

At Betty's Hope, the most reliable collection method, however, comprised deep stone tanks faced with plaster. At Basseterre on St. Kitts, stone cisterns were rectangular, measuring 8 to 9 ft. deep (Hobson 2007:284).

Figure 7.2. Cistern at Betty's Hope.

Compared to those at Basseterre, the cisterns at Betty's Hope were rectangular, larger, and deeper. To the northwest of the great house, four stone cisterns were recorded in the Codrington Papers as 40 ft. long, 14 ft. wide, and 14 ft. deep, holding 586,647 gallons of water (NAAB Codrington Papers:C20/4). Another four cisterns, located to the east of the great house, still visible today, were recorded in the Codrington archives for 1790 as being 20 ft. long, 15 ft. wide, and 10 ft. deep, holding 22,422 gallons of water (NAAB Codrington Papers:C20/4). These four cisterns are not on the 1710 map, but they do appear later on the 1755 map, probably constructed after William Codrington inherited Betty's Hope in 1710. Current measurements of these cisterns come close to the 1790 dimensions, at 18 ft. long, 16.5 ft. wide, and 13.7 ft. deep. The cisterns were probably covered by a roof, which kept out debris, as well as debris filters

Figure 7.3. Walled enclosure, possibly used as an animal pen.

(figure 7.2). Having cisterns was more than just necessary for survival; it was a matter of law, with the imposition of a tax on any plantation owner who did not "possess one or more cisterns" (Lanaghan 1991 [1844]:126).

Another essential part of plantation life was reliance on domestic animals, including pigs, goats, sheep, cattle, chickens, mules, and oxen used for hauling, transporting, and feeding the people who lived on plantations. At Betty's Hope, purchases, sales, and care of domesticates pervade the Codrington accounts. In a letter dated to 1785, specific instructions dictate that "mules should not be under cover at night . . . but they should be penned and locked up" (NAAB Codrington Papers:C15). Animal pens at Betty's Hope are ephemeral, but they are denoted on the old maps. Archaeological site surveys have offered tantalizing clues as to their former locations, which are most likely to the west of the great house, including a large stone wall enclosure surveyed in 2015 (figure 7.3).

A Planting Business

The Caribbean sugar plantation is often described as an agro-industrial complex or "factory in the field," which connotes a machine-like efficiency in management and production (Sheridan 1989:60). Samuel Martin instructs fellow farmers on how to successfully plant, harvest, and process sugar, advocating for different types of fertilizers, such as dung, ashes, seaweed, and cane trash, and promoting crop rotation.

Because sugarcane was planted as a stalk, certain techniques were required to make it grow successfully. Weeding, fertilizing, and watering were just a few of the many activities required to maintain a successful crop. Toward this end, Martin supplied specific instructions in planting cane, known as "cane holing." Martin (1785:18) explained that "the land being well manured and mellowed by fallowing, let it be lined into spaces of four feet distance, and then holed either backwards or sideways, as the manager thinks best, taking particular care that the cane-holes be made square at the bottom, and heaping the banks high, so as to take up little spaces." Cane holing, with its artificial ridges, also saved precious water on drought-prone Antigua (Beinart and Hughs 2007:38). Cane holing was aided by the hoe, the tool of choice. Chris Evans argues (2012:72–73) that the plantation hoe "is best seen as a specifically Atlantic commodity," as different types of hoes were produced for the colonial market to meet the needs of various crops and soils. For the sugar planters, cane holing required a strong blade with a sturdy handle, such as the "Extra-high temper'd" hoe utilized in Antigua by the Tudway estate in Parham (Evans 2012:81–82; Lawrence and Gordon 2013). Hoes and other hand tools were recovered during the excavations of the north windmill (see chapter 14).

At Betty's Hope, fertilizing the cane involved several options. In 1783, plantation manager Joseph Walrond (NAAB Codrington Papers:C15) reports, "We have three modes of manuring. One, by dunging on the spot as above another by throwing the dung into the cane hole, mixing it with a part of the soil and covering it with the remainder. . . . The third mode is to open an hole sufficiently large for two plants nine feet by four and an half for instance, which gives a square of four and an half to each. In the center of this is thrown the dung over which is drawn by earth from each side a pretty high ridge."

The long growing season took anywhere from 16 to 18 months, followed by the highly critical harvest period, during which the cane had to be immediately crushed to extract the cane juice or it would spoil, which is why some planters staggered their harvest period by not having all the cane ripen at once. Martin was very specific in his instructions for the growing season, advising that cane be planted preferably in September or October (Martin 1785:18). This advice was faithfully followed at Betty's

Figure 7.4. View of the Narrows.

Hope, where Christopher Bethell Codrington described it as beginning "in October . . . [but that for planting,] January [is] considered as the most favorable month" (NAAB Codrington Papers:C20/4). At harvest time, the cane was cut into four-foot-long stalks, bundled, and transported by horse or ox-drawn carts to Betty's Hope's two windmills, where the cane was pushed through and crushed by heavy cast-iron rollers (see chapter 12). Following extraction, at Betty's Hope, the juice was boiled down in 16 large caldrons (NAAB Codrington Papers:C2; Carstensen 1993:9; Dunn 1972:195). The cane juice was usually one part water to eight parts sugar, with some lime added. In a May 1715 Codrington correspondence, equipment and supplies for the boiling house include "16 coppers, 7 skimmers, 11 ladels, 4 wagons, 1 cart, 25 yoakes, and 12 chains" (NAAB Codrington Papers:C2).

Although the literature reports the use of copper cauldrons in Caribbean sugar works, at Betty's Hope, only iron kettles have been found at the site and elsewhere on the island (see Meniketti 2006:61–62). After the last kettle boil was completed, the sugar was then placed in wooden flats to be agitated every few hours and to cool. The sugar was then transferred to the curing house, where the crystals were packed into barrels (hogsheads;

hhs) to be drained of molasses for the manufacture of rum. The sugar was further refined by placing the viscid boiled sugarcane juice in earthenware jars known as sugar pots to leach out more molasses from the sugar crystals. The process, known as claying, produced a more refined sugar and left more molasses for rum making. The archaeological evidence in the form of hundreds of clay sugar pot sherds has been recovered at Betty's Hope. Although some planters stopped at this stage, at Betty's Hope, the crystalized sugar was double clayed. In a letter dated to 1723, this process was described as "straining and claying the sugars" (NAAB Codrington Papers:C4).

Shipping the crystallized sugar for the export market was the final step. Large hogsheads were loaded up on carts and taken to nearby wharves, stacked on lighters, and loaded onto ships at St. John's for the Atlantic crossing. Wharves close to Betty's Hope include Parham, Willoughby Bay, and another small wharf, the Narrows (figure 7.4; Lawrence and Gordon 2013:184).

Agricultural Productivity at Betty's Hope

Measuring agricultural productivity at Betty's Hope is challenging for the almost 300 years of its existence. Yet certain trends in both natural disasters and human actions can be linked to production outcomes for much of the eighteenth century. Table 7.1 shows the agricultural output for sugar production from 1701 to 1790 for the Antiguan Codrington estates (NAAB Codrington Papers:C15). For the annual totals within this 83-year period, it should be noted that Betty's Hope would have produced the highest yields since it was the largest holding of the Codrington plantations. For some years, the differences in productivity are quite remarkable. Fluctuations can be attributed to the amount of acreage devoted to cane cultivation, the acquisition of additional land, laws passed by Parliament, hurricanes, droughts, and armed conflict. In correlating the years with significantly lower yields, there are plausible explanations in some instances but not in others.

Hurricanes, earthquakes, and drought top the list of Antigua's natural disasters affecting agricultural output. In addition to destruction of property, hurricanes caused extensive flooding and could lead to loss of infrastructure, such as military forts, as well as famine (Mulcahy 2006:97). Not

Table 7.1. Sugar Produced in Hogsheads (HHs) for the Codrington Estates, Antigua, 1707–1790

Year	HH	Year	HH	Year	HH
1707	43	1733	516	1765	241
1708	135	1734	290	1766	1252
1709	91	1735	461	1767	917
1710	646	1736	408	1768	752
1711	57	1737	120	1769	566
1712	67	1738	510	1770	717
1713	—	1739	548	1771	325
1714	570	1740	439	1772	283
1715	445	1741	450	1773	317
1716	551	1742	361	1774	975
1717	641	1743	526	1775	699
1718	75	1744	570	1776	598
1719	428	1745	405	1777	395
1720	570	1746	512	1778	333
1721	108	1747	108	1779	179
1722	258	1748	548	1780	15
1723	448	1749	529	1781	88
1724	391	1750	530	1782	577
1725	562	1751	449	1783	66
1726	270	1752–1758	N/A	1784	948
1727	329	1759	309	1785	766
1728	612	1760	182	1786	641
1729	710	1761	579	1787	1019
1730	541	1762	592	1788	790
1731	440	1763	271	1789	549
1732	382	1764	469	1790	193

all hurricanes in Antigua's history were recorded, but for the hurricane of 1773, there appears to be a mild correlation with 317 hhs, a lower than normal yield.

Although hurricanes could be devastating, it was drought that truly held planters hostage. Drought was a preoccupation at Betty's Hope, as water was critical to maintaining livestock, for planting, and for domestic and industrial use. One of Antigua's worst drought years occurred in 1779, as reflected in the low output of 179 hhs of sugar. Yet for 1789, during one of Antigua's most severe droughts when 5,000 cattle died, sugar production was robust at 549 hhs.

Drought wreaked havoc in 1833, necessitating shipments of water from nearby islands, and was compounded by a severe earthquake that year (Lanaghan 1991 [1844]:189; Lowe 1951:6). When researching the Codrington Papers, one is struck with the frequency of comments about drought on Antigua, with comments ranging from "The drought is terrible" to "The island is devastated by drought" (Lowe 1951:22), "Worst drought recorded in Antigua" (Lowe 1951:35), "The drought is acute" (Lowe 1951:60), and "A long period of drought is affecting the crop" (Lowe 1951:66). Rainfall charts exist in the Codrington Papers for the periods 1874–1882 and 1892–1893 (NAAB Codrington Papers:E8).

If extreme weather events weren't enough, human actions could influence agricultural endeavors through armed conflict and politics. The low yield for 1760, at 182 hhs, reflects a year of conflict between Britain and France. In 1766, the exceptionally high production of 1,252 hhs correlates to the repeal of the highly objectionable and burdensome Stamp Act, whose repeal may have opened the export flood gates for producers and merchants (O'Shaughnessy 1994).

From 1678 to 1713, as planters asserted their influence and dominance in local assemblies and councils, tensions flared over plunging sugar prices, a rise in the sugar tax, and rising costs in shipping, agents' fees—a period that is referred to as the crisis years (Bayly 2004:41; Dyde 2000:50–51; Dunn 1972:131; Higham 1926:206–207). Additionally, local officials had to be paid, as did managers and skilled workers, and debts had to be settled, some in cash, some in hogsheads of rum and sugar (Hall 2010:59). The Codrington account books consume hundreds of pages of detailed entries relating to these payouts for both Betty's Hope and the other Codrington estates on Antigua.

Another blow came when Parliament imposed a 4.5% export duty on the planters, who surreptitiously escaped paying the tax by loading up vessels in Antigua's hidden bays (Sheridan 1974:48). Although sugar production increased from 1714 to 1720 by 50%, this heady development was sobered by the recession of 1729–1733. Oddly enough, agricultural output seems unaffected by this shift (Sheridan 1974:418–420, 426–431). The years 1739–1748 (except for 1747) demonstrate strong sugar production despite ongoing warfare and were followed by peace and prosperity from 1749 to 1755. The revenues from sugar sales and slave purchases during the 1750s and 1760s, as well as the import of foodstuffs to the Caribbean, indicate a healthy growth in shipping activity and trade within the British

Caribbean (Richardson 1987:766). Sugar production fluctuated wildly, however, during the 1760s, from a low of 182 hhs in 1760 to a whopping 1,252 hhs in 1766.

The most notable contrast in output occurred from 1779 to 1783, with the exception of 1782. During these lean years, output dropped precipitously from 15 hhs in 1780, upticked to 88 hhs in 1781, and then dropped to 66 hhs in 1783, all dangerously low numbers. This can be explained by a series of calamities, including a widespread famine among the enslaved population caused by the American blockade during the Revolutionary War, which prevented food supplies from being delivered to Antigua and Barbuda. The Great Hurricane of 1780, which did not actually land on the island, brought heavy rains, which led to a yellow fever epidemic that may have been responsible for as many as 8,000 deaths (Lowe 1951:6).

Beginning in 1784, production skyrocketed, especially to an impressive output of 1,019 hhs in 1787. That was not to last, however, as 1790 brought with it yet another devastating bout of drought and subsequent disease among the enslaved population. In a letter dated January 6, 1790, Christopher Bethell Codrington remarked that the drought "is the supposed cause of an universal sickness which prevails at this time amongst the negroes all over the island" (NAAB Codrington Papers:C20/1; see Berland et al. 2013). That the sugar yield for that year was only 193 hhs is no coincidence.

The Era of Steam Power

Following Emancipation on August 1, 1834, many of Antigua's plantations were divested and sold. Unlike other planters, however, the Codringtons persevered. Betty's Hope forged ahead, utilizing one of the wonders of the Industrial Revolution: steam power. The Codrington accounts indicate that the move to steam power at Betty's Hope began in 1845 with the purchases and installation of a boiler and machinery (NAAB Codrington Papers:CC462, C48). Oliver (1894–1899:I:175) reports that for 1852, at Betty's Hope, the "two old windmills stand side by side, but are not now used, steam power having superseded them." The accounts also have entries for 1848, with the mention of the purchase of a Cambrian engine, developed in the 1840s (NAAB:CP D1610:A64), sugar refining machinery (NAAB Codrington Papers:A65), and in 1862, purchase of a multi-tubular boiler (NAAB Codrington Papers:A67). The tubular boiler purchase

Figure 7.5. The steam-powered sugar works, Betty's Hope, 1906. Courtesy of the Museum of Antigua and Barbuda.

heralds 1862 as the watershed year for making the full commitment to steam power at Betty's Hope, with a major redesign of the sugar factory. This reconfiguration of the sugar works to steam engines and a tall smokestack chimney marks one of the first of several on Antigua (figure 7.5).

Further improvements were made in 1873, as indicated by the correspondence between plantation manager George Holborow and Sir Gerald Codrington, dated January 11, 1873. In his letter to Sir Gerald, Holborow soberly reported that "after carefully going over the figures, I find it would be necessary either to have . . . new machinery of greater power than we have at present or to remove all the grinding machinery" (NAAB Codrington Papers:C56).

Probably hoping that Codrington would say yes, Holborow carefully crafted his argument by assuring him that after equipment, freight, and installation costs, the £5,463 outlay would pay off by having the "grinding power sufficient to produce 12,000 gallons of [cane] juice per diem or equal to 7 ½ hours" (NAAB Codrington Papers:C56). Holborow prevailed, and Betty's Hope embraced the industrial age. Now a whole new vocabulary took over the Codrington account books with the purchase of a centrifuge engine, engine pump, Cornish boilers, vacuum pans, cane carrier, flues, cast-iron subsiders, and other critical equipment. The scattered remnants of this equipment can be seen at Betty's Hope today, juxtaposing the old and new technologies (figure 7.6). Of course, none of this came without considerable cost. Meniketti (2015:213) was correct in his observation that the characterization of the introduction of steam power on Caribbean sugar plantations as a cost-saving measure is inherently

Figure 7.6. Discarded steam boiler, with older windmill technology in the background, Betty's Hope.

flawed owing to the fact that such endeavors were quite costly, as the Holborow scenario aptly demonstrates.

One other factor may have provided the impetus to modernize: criticism. During his visit to Antigua in the 1850s, John Davy observed that Antigua was behind the times, as the "old methods seemed to be generally kept in use; the windmill for crushing the cane, rather than the steam engine, with a very few exceptions; the open boilers of taiches rather than the vacuum pan, or other more refined methods for evaporating the juices" (Davy 1971 [1854]:402). Davy, however, does credit agriculture on Antigua as "slowly improving, becoming more skilled" and "the hoe has ceased to be, what it was so long, the only implement employed in tillage" (1971 [1854]:399).

Tossing the hoe aside, in 1863, Holborow invested in a steam-powered plow for Betty's Hope, thinking that it could cut field labor costs (NAAB Codrington Papers:E7). Unfortunately, the wonders of this modern piece of machinery failed to deliver. Holborow bitterly complained that on its inaugural day, "the screws that hold the plough together have been all broken." To compound matters, the soil was "as hard as brick," wreaking havoc with the machinery. Holborow lamented that the plow "never ought to have been put into it." Furthermore, Holborow suggested that the plow would be much improved if run on "two engines of ten horsepower each" (NAAB Codrington Papers:C52).

Innovation or Status Quo?

In investigating the long trajectory of Caribbean cane farming, the question arises as to whether the planters were truly innovative in their approaches or just plodding along. Paquette and Engerman (1996:6) maintain that for the period 1650–1750, the plantation system reflected "an economic enterprise of the first order that required a sophisticated integration of production and processing and an intensive use of the factors of production." On the other hand, Sheridan (1960:126–127; 1989) found very little in the way of innovation prior to 1800, which he attributes to a system based on enslaved labor, which did not motivate innovative thinking, with the exception of Samuel Martin, who "was an innovator in his own right" (Sheridan 1960:139; see Hicks 2007:44).

Looking back at the long trajectory of agricultural practices at Betty's Hope, Justin Roberts (2013:37–38) reminds us that technological innovations are easy to overlook on Caribbean sugar plantations because of the reliance on hand tools, work being done in slave gangs, and on wind power until the advent of steam, particularly if technology only refers to machinery. Innovation, however, can be measured in other ways on Caribbean sugar plantations.

The Codrington Papers reinforce the notion that Betty's Hope's owners and managers were unquestionably interested in improving plantation operations. One way to think about innovation relates to cutting costs and increasing or improving self-sufficiency in such things as animal husbandry (Roberts 2013:39). The Codringtons did this exceedingly well by raising some of their food and livestock on Barbuda (see chapter 4). Galloway (1985:335) states that as rational actors, planters were well aware of agricultural innovations but adopted them "when it made good economic sense to do so." Galloway (1985:337) credits three factors that effected the adoption of innovative agricultural practices: scarcity in natural resources that required resourceful solutions, the overall cost-benefit ratio, and market competition.

This can be seen at Betty's Hope in a number of ways. First, to reap profits, the overall scale of sugar production had to be increased, which the Codringtons and their managers accomplished through increasing their land holdings and devoting more acreage to cane cultivation. Another factor was maximizing mill technology. To meet the demands of daily capacity during harvest, often more than one mill was required. Betty's Hope was one of the few plantations in the Caribbean to maximize the power of two windmills.

Design in mill technology was also critical. Generally, windmills of the early eighteenth century relied on three vertical iron rollers, which was inefficient in that it required two men to push the cane stalks through the rollers. Taking into consideration that each cane stalk had to be crushed twice to extract as much juice as possible, this system was highly inefficient. To maximize efficiency, Betty's Hope switched to a new system that employed three horizontally positioned rollers, which required only one cane feeder, with an increased extraction rate of 80% (Museum of Antigua and Barbuda 2005). Galloway (1985:342) also credits the use of cane trash (bagasse) to provide fuel for carrying out the boiling of the cane juice, mainly because islands like Antigua had been deforested, so the natural resource of wood as fuel was unavailable, thereby creating the need for a resourceful solution.

As Roberts states (2013:279–280), the "spirit" of improvement permeated the plantation complex for the latter part of the eighteenth century, because there was significant economic incentive to do so. This is evident at Betty's Hope. Yet what has been missing from the discussion are the important contributions of enslaved Africans in any experimentation, inventions, or innovations that occurred on Caribbean sugar plantations. Recent scholarship is beginning to recognize and acknowledge the contributions of enslaved Africans to the plantation enterprise. We know that it took great expertise and technological sophistication to crush cane and process cane juice, and that expertise can stimulate new ideas and ingenuity in tackling challenges. Skilled enslaved Africans also had knowledge and expertise in many other related areas, such as carpentry, cooperage, rum distilling, and ways to improve soil fertility. In fact, planters were probably relying more on their enslaved labor for their versatility and ability to develop cost-saving and labor-efficient solutions (Roberts 2013:206, 236).

The lack of voice in the written record may seem to stymie our understanding of what these contributions were. However, one clue may be to examine accounts and interviews recorded in the earlier part of the twentieth century of descendant communities. Sir Eustace Francis, whose father was a manager at Betty's Hope in 1944, recalled the "ingenuity and knowledge" of the workers in their planning and carrying out controlled field burns to clear the land (Lawrence and Gordon 2013:186). Although ethnographic accounts for more than a century after Emancipation are not necessarily reflective of past behavior, such accounts can offer a glimpse into possible resourceful practices that may have existed in pre-Emancipation times. Can archaeology provide clues? This topic is definitely an area ripe for research.

8

The Landscape Legacies of Sugar and Rum

An Historical-Ecological Perspective from Betty's Hope

SUZANNA M. PRATT, E. CHRISTIAN WELLS,
AND ANTHONY R. TRICARICO

The environmental legacies of sugarcane cultivation for sugar and rum at Betty's Hope Plantation are visible on the contemporary landscape through differential impacts to soil quality. Our research utilized the erosion-productivity impact calculator (EPIC) to simulate different fertility and erosion characteristics of local soils based on contemporary environmental parameters and on information recorded in the Codrington Papers, dating from 1700 to 1944. The resulting models were evaluated against soil data collected from former plantation fields around Betty's Hope in 2014 and 2015, which recorded contemporary levels of organic matter and plant macronutrients. We found that agriculturally abandoned landforms in the area are highly degraded (low fertility, high erosion), while landforms with active and recent farming activity have been less affected by degradation. These results support emerging research in historical ecology (Fisher 2005; Håkansson and Widgren 2014; Wells at al. 2013), suggesting that the historical abandonment of highly engineered landscapes can create environmental conditions conducive to significant and lasting soil erosion and loss of soil fertility. This finding challenges previous assessments of plantation soilscapes in the Caribbean (Dillman 2015; Meniketti 2015; Watts 1987), which are believed to be uniformly and permanently degraded from monocropping and sustained, long-term farming. We contend that while plantation agriculture had major impacts on soil health at Betty's Hope (see Fox 2016; Tricarico 2019; Wells et al. 2017; Wells et al. 2018a; Wells et al. 2018b), the somewhat recent abandonment of fields that were planted and maintained over hundreds of years also precipitated large-scale erosional events in the area. Plantation agriculture and its cessation at Betty's Hope are the distal and proximate causes, respectively, of soil degradation.

Antiguan Sugarcane Cultivation in Historical Context

There is no mention of sugarcane cultivation in Antigua before 1655 (Dyde 2000:20), but once established in the late seventeenth century, the industry dominated local agriculture. Antigua became one of the most productive sugar islands in the West Indies. By 1750, nearly all of the virgin timber had been cleared from the island for sugar production (Watts 1987:340). In 1764, Antigua had over 300 sugar plantations, each an average of approximately 200 acres, operated by about 100 enslaved Africans (Watts 1987:340).

An increasing labor supply and a limited amount of arable land contributed to the development of intensive cultivation methods on Antigua as plantation owners sought to extract the most profit from their land (Watts 1987). Compounding the inherent difficulty of growing a labor-intensive crop was Antigua's lack of freshwater. Emmanuel Brown, in his 1750 map of Antigua, noted "Antigua is a fine Island, tho' it has not one Single Spring of Water in it: So that in times of drouth when all their Ponds are dry, and their Cisterns almost empty, they are obliged to fetch their fresh water from Montserrat . . . or . . . Guardaloup, a Neighboring French Island" (Dyde 2000:3). Sheridan (1960:127–133) notes that plantation owners, particularly Samuel Martin of Greencastle Estate in Antigua, actively tried to combat reduced yields caused by land degradation by employing a number of novel techniques, such as fallowing, fertilizing with manure, and tilling. Sugarcane production subsequently skyrocketed, rising from an annual average of 4,900 tons in the 1710s to 9,200 tons in the 1760s (Sheridan 1960:127). In 1724, the governor of Antigua reported, "No land on the island remained unplanted" (Sheridan 1960:127). From the period of 1710 to 1750, Antigua and other Leeward Islands (St. Kitts, Nevis, and Montserrat) collectively became the greatest sugar colonies in the region (Watts 1987:232). This economic prosperity is reflected in demographic changes on the island. The population of enslaved laborers (about 12,500 in 1713) increased steadily, reaching almost 375,000 in the mid-1700s, though the number of enslaved Africans imported was never enough to satisfy the demand for labor (Dyde 2000:65). At the height of

Antigua's sugarcane industry in the late eighteenth century, over 90% of the island was devoted to agriculture (United Nations 2005:22).

The varying quality of Antigua's soils, combined with uncertain rainfall, created conditions unfavorable for sustained high-yield monoculture (Midgett 1984:34). The island experienced a decline in production toward the end of the eighteenth century due to a combination of factors, including soil exhaustion, crop disease, drought, absentee plantation owners, competition from other Caribbean islands, and reduced returns of increasing investment of capital and labor in a limited amount of land (Ragatz 1928; Sheridan 1960:127; Smith 2005:199; Ward 1978:198). From the 1750s until the outbreak of the American War of Independence, the output of sugar from Antigua and the rest of the Leeward Islands declined. The slump in production coincides with a general pattern of decline across sugar plantations in the Eastern Caribbean during this period (Lowes 1994a:8; Watts 1987:315).

In the latter half of the eighteenth century and in the first half of the nineteenth century, Antiguan planters were devoted to large-scale sugar production at a time when prices were falling and increased competition favored the most economical producer. The primary challenge for the Antiguan sugar industry was not inadequate soil nutrients but the island's susceptibility to repeated cycles of drought, some lasting two or three years (Auchinleck 1956). Lowes (1994a:10) examined island-wide sugar yields from 1820 to 1870 and found that sugar production "swung wildly from one year to the next," but that the reported yields display a steady decline. The concerns about drought and unpredictable water supply are evident at Betty's Hope with the addition of four enormous tanks for water collection by the late seventeenth century.

Difficulties in maintaining a high crop yield led to the successive abandonment of sugarcane operations. By 1943, when most Antiguan sugar estates were consolidated into the Antigua Syndicate Estates, Ltd., the island's sugar acreage had already been reduced by half (Midgett 1985:34). Production continued to decline, with less than 5,000 tons per year by 1970 (Midgett 1985:35). The Antiguan sugar industry all but ended in 1972, when the last sugar refinery closed and sugar production fell to "negligible levels" (Dyde 2000:278; Weaver 1988:321). Today, sugarcane mostly grows in relic patches along roadsides and in abandoned agricultural fields. Many of the fields surrounding Betty's Hope are actively used for a wide variety of purposes, including small-scale farming (papaya, peppers, tomatoes, beans), beekeeping, cattle ranching, and urban development (both residential and commercial).

Methods

EPIC Simulation

The erosion-productivity impact calculator is a mathematical model for simulating erosion, crop production, and related physical processes for different agricultural regions, allowing the input of data pertaining to specific factors for a given location and scale per crop type and the output of reliable erosion and productivity predictions over time (Easterling et al. 1992:18; Wingard and Hayes 2013:xvi). The model was created in 1981 by the US Department of Agriculture (Easterling et al. 1992:18; Gassman et al. 2005:2; Williams et al. 1984) to assess the effects of soil erosion on productivity and predict the effects of land management decisions on soil, water, nutrient, and pesticide movements and their combined impacts on soil loss, water quality, and crop yields for areas with homogeneous soils and management (Blacklands Research and Extension Center 2014). The model simulates a hydrologic land-use unit, which is often a field, farm, or small watershed that is homogenous in climate, soil, land use, and topography. To acknowledge the broadened range of applications of the simulation, the EPIC acronym now stands for "erosion policy impact climate" (Gassman et al. 2005:2).

While EPIC has mostly been used to predict soil erosion and productivity, we used it to retrodict these characteristics for Betty's Hope based on historical archival data from the estate, following recent examples using historical (Huang et al. 2006) and archaeological (Wingard 2013) data. The major advantage of the EPIC model for our research is that it can predict crop yields based on specific geographic, climate, and crop data, allowing for the simulation of agricultural impacts over long periods. The major drawback of the model is that it is designed for modern mechanized agriculture (Texas A&M AgriLife Research 2014) that cannot necessarily be measured or estimated for cultivation in the past. Additionally, the exact climate and soil conditions and the area of acreage under cultivation at any one time are largely unknown for Betty's Hope. However, for the purposes of this research, the EPIC model serves to create a hypothetical simulation of crop yields under ideal conditions that can serve as a baseline for better understanding erosion and productivity associated with the plantation's fields.

The model incorporates nine discrete categories of variables: hydrology, weather, erosion, nutrients, soil temperature, plant growth, tillage, plant environment control, and economics (Easterling et al. 1992:18; Gassman et al. 2005:4; Williams 1990:421). In this study, the EPIC model was created using publicly available data for each of the nine categories of variables. Weather data were obtained from WeatherSpark, which generates interactive weather history graphs based on data obtained from the National Oceanic and Atmospheric Administration, the Norwegian Meteorological Institute, and World Weather Online (WeatherSpark 2014). The Antiguan weather report documents the typical weather in Antigua based on observations recorded between 1974 and 2012. The average monthly values for precipitation, temperature, humidity, and wind speed, plus calculated standard deviations for monthly minimum and maximum temperatures and standard deviations and skew coefficients for monthly precipitation, were input into the model to generate weather patterns.

Soil data for the EPIC model were based on values obtained during laboratory analyses of the soil samples recovered from Betty's Hope (see Pratt 2015; Wells et al. 2018b). A deep auger probe located in a fallow field was used to recreate a hypothetical soil profile. The changes in mineral and particle concentration down this profile follow the expected patterns for a normal soil profile for a cultivated field. In order to simulate a simple agricultural field using EPIC, the values for pH, organic carbon, calcium carbonate, and soil texture were used from this probe to build three soil horizons. Since sugarcane growth is optimal in well-drained, deep, loamy soil with a bulk density of 1.1 to 1.2 g/cm^3 (Bakker 1999:16), the bulk density for each layer in the EPIC model was assigned a value of 1.2 g/cm^3.

The inclusion of agricultural operations into the EPIC model proved challenging. The EPIC model is designed to simulate modern farming activities; however, large-scale mechanized agricultural techniques did not exist in Antigua prior to the twentieth century. As a result, we developed a simple agricultural operation with three steps: planting of crops in a row, harvesting the crop, and termination of the crops. No operations for tillage or fertilization were included in order to simulate annual crop yields if human activity was kept constant. This disregards the potential effects of investments of capital, including liming, manuring, fertilization, construction of cane holes, and tilling. Finally, sugarcane takes approximately

18 months to mature before it is harvested. In Antigua and the rest of the West Indies, sugarcane was typically planted between August and November, allowed to grow for a year and a half, and harvested between January and June (Watts 1987:176). To account for the multiyear growth period, the duration of the agricultural operations was set to span three years. Consequently, these settings produced crop yield estimates for every third year. These estimates represent the hypothetical amount of sugarcane produced per acre of land per planting/harvesting cycle.

Historical Data

For historical data about sugarcane yields, we examined archival documents from the Codrington family (who owned and operated Betty's Hope from 1674 until 1944). Among the collections are four documents recording sugar sales: "Annual Statements of Total Sugar Crop 1801–1838, with Calculations of Profits from Antigua and Barbuda 1707–1830, etc." (SFU 1838); "Sugar Accounts of C. B. Codrington, with M. Trattle, merchant 1802–1807" (SFU 1807); "Sugar Accounts of C. B. Codrington, with M. Trattle, merchant 1807–1813" (SFU 1813); and "Account Sales Book of C. B. Codrington for Sugar and Wool 1824–1828" (SFU 1828).

The first document provides a record of the annual sugar produced by all Codrington estates in Antigua, including Betty's Hope and three other plantations: Cotton, Garden, and New Work. The other three documents detail the total amount of sugar or rum sold in a single transaction, as well as the date of the sale, the transfer of goods, the taxes and costs associated with shipping and customs, and the net profit of the sale in British pounds (figure 8.1). These records provide exact amounts of sugar and rum sales in quantities of hogsheads (wooden barrels containing 63 US gallons), tierces (smaller wooden barrels containing 42 US gallons), and puncheons (large barrels containing 84 US gallons). While the documents record the total amount of sugar produced and sold from all Codrington estates in Antigua, the information is useful for deriving a proxy measure for the crop yield at Betty's Hope.

The quantities of sugar or rum recorded in each document were totaled and converted to compatible units. For "Annual Statements of Total Sugar Crop 1801–1838, with Calculations of Profits from Antigua and Barbuda 1707–1830" (SFU 1838), annual crop yield (in hogsheads) was provided in a list. For the sales records with Marmaduke Trattle, the amounts of

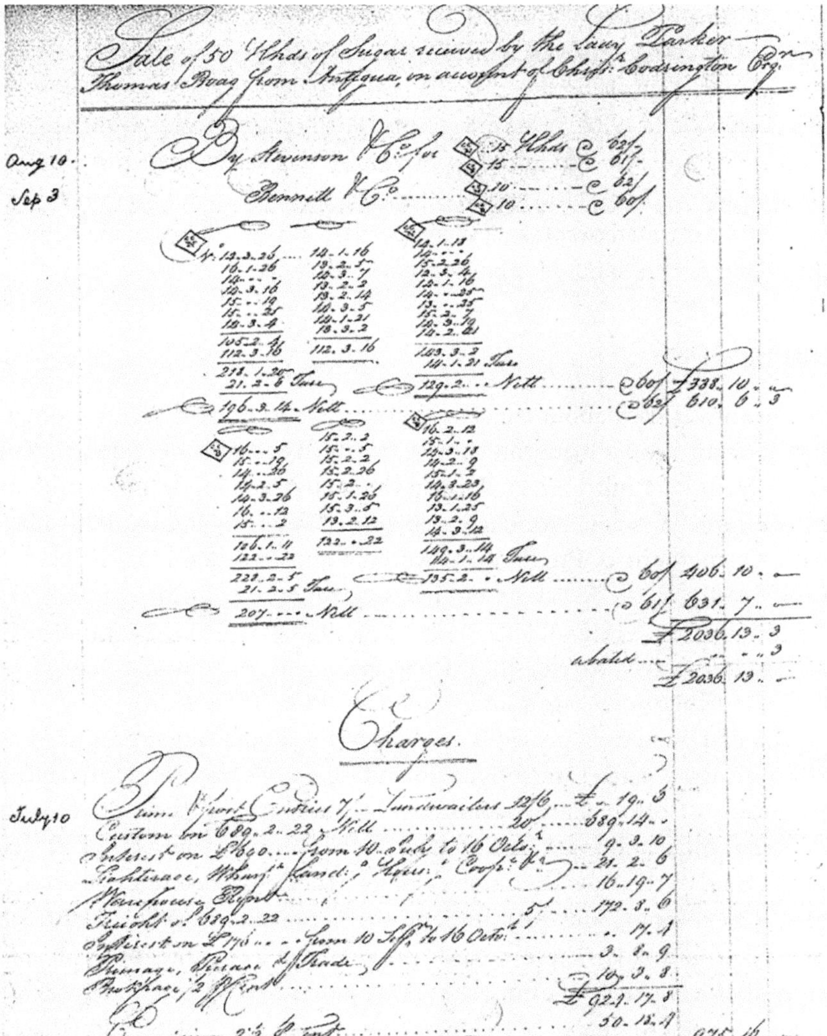

Sale of 50 Hhds of Sugar received by the Lady Parker
Thomas Boag from Antigua, on account of Chris. Codrington Esq.

Aug 10. By Stevenson & Co. for 15 Hhds @ 62/
15 @ 61/-
Sep 3 Bennett & Co. 10 @ 62/
10 @ 60/

129.2.. Nett @ 60/ £338. 10. –
196.3.14. Nett @ 62/ 610. 6. 3
135.2.. Nett @ 60/ 406. 10. –
207.... Nett @ 61/ 631. 7. –
£2036. 13. 3
abated 3
£2036. 13. –

Charges

July 10 Prim. Report Entries 7/- Landwaiters 12/6 £ – 19. 6
Custom on 689.2.22 Nett 20/ 689. 14. –
Interest on £690 ... from 10 July to 16 Octr. 9. 3. 10
Lighterage, Wharfage, Land., Hous., Coop. &c. 21. 2. 6
Warehouse Rent 16. 19. 7
Freight on 689.2.22 5/ 172. 8. 6
Interest on £170 ... from 10 Sept to 16 Octr. – 17. 4
Primage, Petty Charges &c. 3. 8. 9
Brokerage, 1/2 per Cent 10. 3. 8
£924. 17. 8
Commission 2 1/2 per Cent 50. 18. 4
975. 16. –
1802 Novr. 6 To his Account Current for Nett Proceeds £1060. 17. –
2. 1. 9
1062. 18. 9

Figure 8.1. A sugar sales record from 1802 titled "Sale of 50 hhds of Sugar received by the Lady Parker. Thomas Boag from Antigua, on account of Chris. Codrington Esq." from Sugar Accounts of C. B. Codrington, with M. Trattle, merchant 1802–1807 (SFU D1610 A60/1).

sugar and rum recorded in hogsheads, tierces, and puncheons sold per year were totaled and then converted to hogsheads (one hogshead is equal to 1.5 tierces and .75 puncheon). This allowed us to total the number of hogsheads per year produced by all Codrington estates in Antigua from 1707–1751, 1760–1779, 1782–1790, 1802–1814, 1817–1826, and 1828. The gaps in the record are due to the lack of records for these years or the poor quality of the microfilm that rendered the documents illegible.

In the mid-nineteenth century, sugar hogsheads from the West Indies contained 1,456–1,792 pounds of dry goods, or the equivalent of 63 gallons of liquid (Waterston 1859). To convert hogsheads to tons, we multiplied the number of hogsheads by 1,456 and divided by 2,000. These figures represent the amounts of sugar products produced by the Codrington estates, serving as a proxy for raw sugarcane reaped from the fields. The ratio of raw sugarcane to processed sugar, known as the recovery rate, indicates the quantity of cane required to produce a ton of crystalline sugar (Abbott 1964:12). In 1940, Antigua's ratio of tons of sugarcane to tons of sugar was 8.58 to 1.00. However, this measure of productivity was calculated for the end of Antigua's tenure as a successful sugar island, and Abbott (1964:16) notes that Antigua was "the most inefficient producer of all the West Indian islands" at this time. We multiplied the amounts of sugar products sold (in tons) by 8.58 to determine approximate annual crop yields in tons. Although the recovery rate of tons cane to tons sugar was not constant over time, the multiplication by the recovery rate from 1940 does not affect the variation in calculated crop yields from year to year. Using the reported quantities of sugar products sold from 1707 to 1828, we can extrapolate historical crop yields (in tons) and assess the trends in annual yields over time.

Geoarchaeological Assessment

To evaluate the results of the EPIC model and the historical data, we conducted a geoarchaeological study of the area surrounding Betty's Hope. The basin encompassing Betty's Hope is composed of Plio-Pleistocene coral limestone, which contributes calcareous and kaolinitic clay soils to the region. The basin is approximately 22 km^2 and contains a series of overlapping and interfingering alluvial terraces and colluvial fans, drained by Mercer's Creek on the north side of Betty's Hope and Ayer's Creek on the south. Agricultural fields surrounding Betty's Hope are mainly composed of broad, flat alluvial terraces, with residential settlements located on higher (older) terraces. Our geoarchaeological study included geomorphological survey with soil sampling and analysis, following procedures outlined by Wells and colleagues (2013).

We conducted a full-scale pedestrian survey of the entire basin, identifying and mapping prominent landforms and other features (e.g., current land use, archaeological remains) in the field. These maps were digitized and georeferenced using ArcMap (v. 10.3; ESRI, Redlands, CA, USA) and were overlaid with a map of Betty's Hope estate from 1755 (NAAB Codrington Papers:P4). This allowed us to match observed geomorphological landforms with the precise locations of former plantation fields. We used this conjoined map to then determine the locations of soil profile samples, which we obtained in the field using a standard stainless-steel bucket auger. In all, we sampled 32 profiles from four distinct landforms. For each probe, we examined the soil sequence in the field (from topsoil to bedrock), taking notes and photographs and making preliminary assessments of color, texture, structure, and other visible variables. We selected a ca. 50 g bulk sample from each identified horizon within each probe and placed these into sterilized Whirlpak bags for transportation back to our laboratory at the University of South Florida. In total, we collected 195 soil samples.

Soil samples were analyzed for color (Munsell, moist with natural light), texture (pipette method), pH (glass electrode, 1:1 mixture), and organic matter (loss-on-ignition, 360° × 2 h). We also used a portable X-ray fluorescence spectrometer (Bruker Tracer III-SD) to determine the major and minor elemental composition of individual soils. Our assays included 19 elements: Na, Al, K, Ba, Ti, Mn, Fe, Co, Ni, Cu, Zn, Ga, Pb, Th, Rb, Sr, Y, Zr, and Nb. Sample preparation consisted of powdering 5 g of each sample with an agate mortar. Each pulverized sample was placed into a clean Whirlpak bag and then against the analytical window of the instrument. The analysis of a sample through the plastic barrier of the bag has no effect on sample readings since heavier elements fluoresce higher energy X-rays and thus experience little alteration while traversing through a thin plastic barrier (EDAX Smart Insight 2014). This was tested by analyzing one soil sample four times, twice through the plastic bag and twice without the plastic bag. The results for all four analyses were highly similar, indicating that the presence of the plastic bag had no effect on pXRF measurements. Elemental data were collected without a vacuum at 40 kV/11μA for 120 seconds. Each sample was analyzed twice, with the instrument's beam directed at a different part of the sample. Our analysis of the data uses average values between the two readings.

Results

EPIC Simulation

The EPIC model produced annual values for sugarcane yield (tons/acre), humus mineralization (pounds/acre), nitrification (pounds/acre), evapotranspiration (inches), phosphate mineralization (pounds/acre), and water erosion. The EPIC model was given parameters for Antiguan weather and soil and set to simulate a 1.5-year sugarcane crop cycle from 1650 to 2049. Due to sugarcane's 18-month maturation cycle, EPIC hypothesized crop yields per acre for every third year. While Betty's Hope likely had multiple staggered crops producing sugarcane annually, the EPIC model provides a proxy for the overall simulated crop yield. Since weather parameters, soil parameters, and crop operations were held constant for the entirety of the simulation, we expected that the EPIC simulation would produce crop yields that gradually declined over time. This expectation is informed by the suggestions of previous researchers who observed that continuous monocropping causes a deterioration of the physical and chemical properties of soil and thus smaller yields over time (Abbott 1964:1; Garside et al. 2001:16; Meyer et al. 1996; Ragatz 1928; Ward 1978:198). Lowes's (1994a:10) analysis of sugar production trends in the latter half of the nineteenth century suggests a steady decline of crop yields, despite yearly variation.

A line graph comparing simulated and historical crop yields shows initial overlap and then key periods of divergence over time (figure 8.2). To represent the trends in the simulated and historical yields visually, the EPIC-simulated yield was multiplied by 500 (to represent tons/500 acres) in order to scale the simulated and historical yields similarly. In order to smooth the variation of yearly production and assess broader trends, we computed six-year moving averages for both the historically recorded sugar yields and the EPIC-simulated yields. This follows the strategy employed by Lowes (1994a). The use of moving averages also bridges analytical gaps created by periods when data are missing from the historical records and the period in which the EPIC simulation and historical data

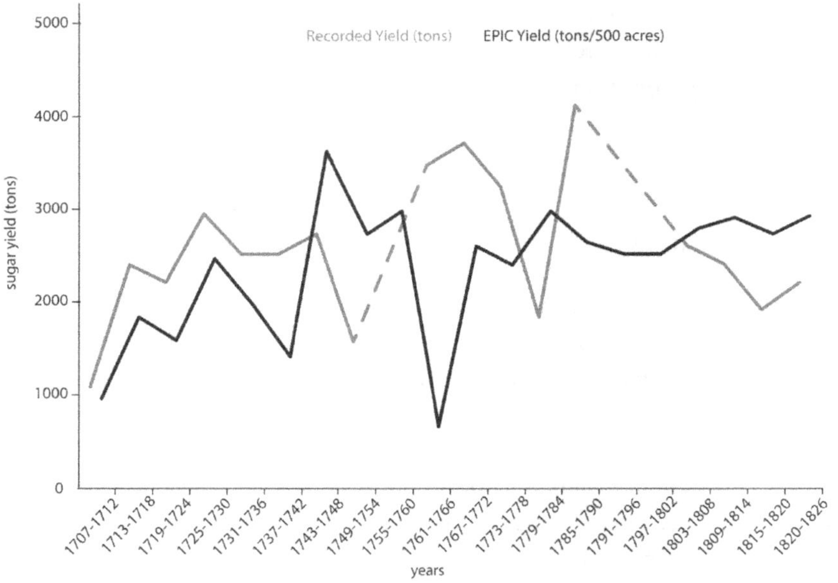

Figure 8.2. Line graph of the changes in the six-year moving average of the historically recorded and EPIC-simulated sugarcane yields.

vary. The comparison of the historical production and the simulated crop yield allows us to assess whether trends in the historical yield are consistent with those of the simulation. However, because the exact amount of land cultivated by the Codrington estates is generally unknown, the changes in historically recorded sugar yields may be due to changes in the size of the sugarcane fields associated with Betty's Hope.

It is immediately apparent that simulated crop yields decline dramatically from 1652 to 1658 (the first and third crops). Although variable from crop to crop, the yield then declines steadily until 1745. From 1745 to 1766, crop yields are extremely variable, with a hypothetical yield of .84 ton/acre in 1745, 14.35 tons/acre in 1748, .05 ton/acre in 1751, 10.98 tons/acre in 1745, .05 ton/acre in 1757, 11.0 tons/acre in 1760, .15 ton/acre in 1763, and 2.57 tons/acre in 1766. Following this period of dramatic variation, the simulated yields become more consistent, ranging between 4 and 7 tons/acre from 1769 to 2048.

The comparison of the six-year moving averages reveals an anomalous decrease in historical yields from 1755 to 1760 and a similar decrease in the EPIC-simulated yield a few years later, from 1761 to 1766. Aside from the unexpected decrease, the historical yield increases steadily from 1707 to the period of 1797–1802, then decreases more sharply until 1826. This period coincides with a known expansion of the sugar trade and acreage for planting; there is likely an increase in the purchase of enslaved Africans to cultivate increased acreage to meet the demands of the market (Abbott 1964:1; Campbell et al. 1992; Garside et al. 2001:16; Meyer et al. 1996; Ragatz 1928; Sheridan 1960:135; Ward 1978:198). The simulated yields are more variable, with a slight increase from 1707 until the unexpected decrease in 1761–1766 (coinciding with part of the period of dramatic variation in the EPIC simulation), then is relatively stable until 1826. This period of 1755–1760 also coincides with an absence of data in the historical records (from 1752 to 1759). The unexpected declines in sugar yields in both the simulation and the historical data likely do not reflect an actual decline in crop yields during this time as the simulated yields are unexpectedly variable and historical records are missing. Additionally, the Antiguan sugar industry reached its peak during this time, thus a dramatic decline in sugar yields in the 1750s is not expected. The divergence of the historic yields also suggests that human inputs into the agricultural landscape, such as fertilization, cane holing (the process of digging a hole to plant the cane, then gradually infilling with debris or manure as the cane grows), drainage, liming, and irrigation and drainage features, may have prevented soil exhaustion and erosion simulated by the declining EPIC yields.

In addition to examining the simulated crop yields at Betty's Hope over time, we considered simulated rates of nitrification and phosphorus mineralization for sugarcane (figure 8.3). Nitrification refers to the oxidation of ammonia/ammonium by nitrite-oxidizing bacteria to nitrate (NO_3^-), which is a plant-available form of nitrogen. Phosphorus mineralization refers to the microbial conversion of organic forms of phosphorus to orthophosphates ($H_2PO_4^-$ or HPO_4^{2-}). Both nitrogen (N) and phosphorous (P) are key mineral nutrients for plant growth. Decreases in nitrification and P mineralization, and thus decreases in plant-available forms of N and P, can result in reduced crop yields and poor soil health, which can be counteracted with the application of fertilizers, manures, other soil additives, or fallowing. Figure 8.3 suggests that N and P would have become major challenges to sugarcane production at Betty's Hope beginning around 1700, with significant drop-off by the 1750s. This pattern is consistent with the actions taken by Antiguan farmers in the 1720s, when they introduced an intensive process of fertilizing, manuring, and digging

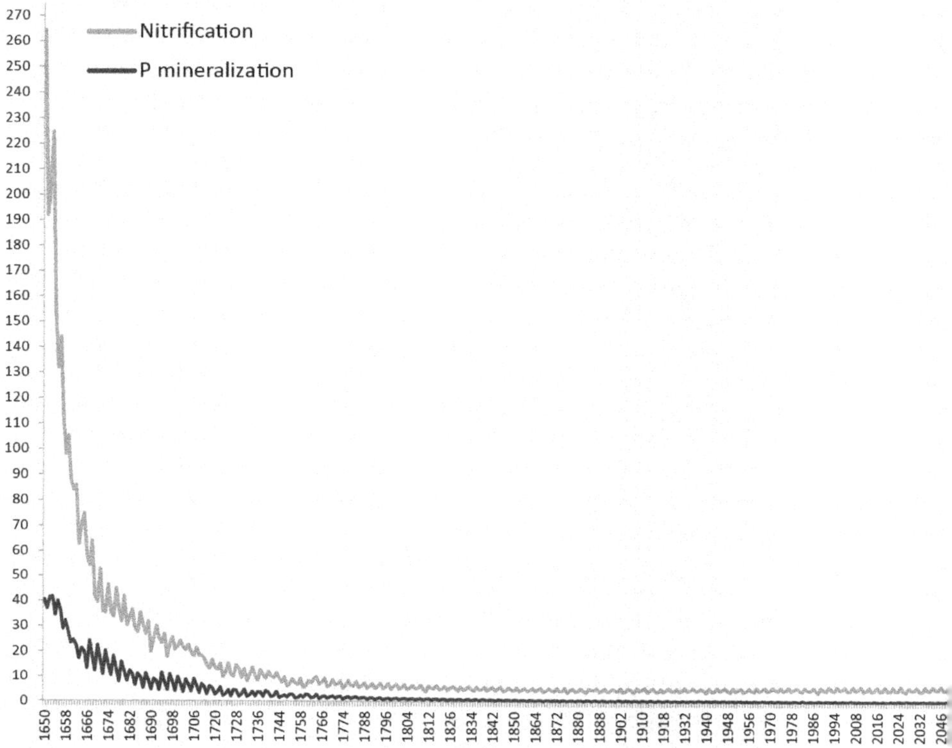

Figure 8.3. Line graph of EPIC-simulated rates of nitrification and phosphorus mineralization for sugarcane at Betty's Hope, 1650–2046. The graph shows a sharp decline over the first 50 years, followed by a gradual leveling off at very low levels by the early 1700s.

cane holes to concentrate fertilizers near the base of the cane (Martin 1784:271–272). This practice demanded constant attention and maintenance, and where ignored, soil health declined (Sheridan 1960:134).

Historical Sugar Yields

The first historical records of Codrington crop yields in this analysis appeared in 1707. Despite the dearth of records for this decade, the yields increased, paralleling the historical development of the West Indian sugarcane industry when Antigua became the lead regional producer. The War of Spanish Succession, a major European political conflict, took place during this decade, lasting from 1702 to 1713. However, despite the turmoil in Europe and the absenteeism of the Codringtons, sugar was a profitable and increasing industry in Antigua.

Codrington crop yields declined in 1711 and then increased slightly in 1712. No record of the 1713 crop is available, but yields remained relatively high for the next few years. By 1715, Betty's Hope was capable of industrial sugar production; the estate had 16 copper kettles for boiling cane juice and four stills for distilling rum (Goudge 2017). Sugar yields continued to vary in the second half of the decade, dropping in 1718 and then increasing at the end of the decade. The six-year moving average of sugar yields reveals a trend of increase in the first half of the decade, then a slight decrease in the second half. Despite the variation in annual yields, the Codrington estates produced roughly the same amount of sugar at the beginning and end of the 1710s.

No major climatic events or events related to the Codrington estates are known for the 1720s or 1730s. Sugar yields were variable once again, though less dramatically than in the previous decade, indicating consistent sugar production. The six-year moving average of sugar yields indicates a slight increase in the first half and a slight decrease in the second half of the decade, though there is a small net increase in average sugar yields.

In the 1730s, yields declined, with small amounts of variation at the beginning of the decade, and a more dramatic decrease occurred in 1737. This coincides with an attempted rebellion by enslaved Africans in Antigua in 1736 (Brown 2002:104). This rebellion and the associated social turmoil are accompanied by a sharp decline in sugarcane yields. The year 1737 also marked a time of expansion at Betty's Hope. The construction of the north windmill tower is a sign of expanding production and the increased need to supply more power to process sugarcane into exportable products. This suggests that sugarcane yields were increasing or expected to increase and reflects Antigua's role as the dominant sugar island of the region, devoted to growing a single crop. Sugarcane yields increase in the last few years of the decade. The six-year moving averages of sugar yields indicate that average yields were relatively stable during this period; this is consistent with the two previous decades. During the first half of the eighteenth century, sugar production was a growing and profitable industry in Antigua and the economic dominance of a single crop is reflected at Betty's Hope in the expansion of the sugar production capacity and in consistent sugarcane yields (Fox 2016).

In the 1740s, sugarcane yields were steady, with very slight annual variation at the beginning of the decade. A decline is observed in 1747, and then yields increased to approximately the same levels as earlier in the decade. The majority of the War of Jenkins' Ear (and the War of Austrian Succession that subsumed it, 1739–1748) occurred in this decade, creating an atmosphere of conflict and uncertainty in the Caribbean. Despite Spanish and British forces at odds in the region, Betty's Hope appears to have been largely unaffected by these conflicts; the Codrington estates produced 439 hogsheads of sugar in 1740. This success in sugar production is reflected in yet another expansion in 1741, when the Betty's Hope Boiling House acquired five additional copper boiling vats. As with the expansion in the 1730s, this construction reflects the increasing sugarcane processing capacity at Betty's Hope, speaking to the success and profitability of sugarcane cultivation. Despite the decline in 1747, the six-year moving averages indicate an overall increase in sugarcane yields during the 1740s. As Antigua was nearing the peak of its production, the increase in crop yields and expansion at the Codrington estates again parallels the historic expansion and dominance of sugarcane in the region.

Antiguan sugarcane reached its peak in 1753. However, records of the Codrington sugarcane yields are missing for this decade. An unusually dry year was recorded in 1752, likely affecting sugarcane yields across the island, though there is no way to determine whether this event is associated with a decline in Codrington sugar yields. The success of Betty's Hope can be inferred, however, as historical records indicate that the plantation had 227 enslaved laborers in 1751 (see chapter 10). Additionally, the Seven Years' War began in 1756, ushering in a phase of economic prosperity for British planters as British naval forces ruled the seas, reducing French competition and increasing access to British markets (Watts 1987:276).

The Seven Years' War ended in 1763, marking the conclusion of a highly successful and profitable period for British colonial sugar (Ragatz 1928:111). The wealthy wartime years marked the beginning of the end for British planters' wealth (Watts 1987:276). From the 1750s until the beginning of the American War of Independence, the output of sugar from the Leeward Islands became increasingly difficult to maintain at previous levels, and the output from Antigua fell (Watts 1987:315). Historical sugar yields from the Codrington plantation were variable in the beginning of this decade, but overall appear relatively consistent. A notable spike in yields occurred in 1766, then yields returned to levels similar to those of the beginning of the decade. The six-year moving averages for this decade show much higher yields than in previous decades, with an overall trend of increase. From these crop yields, it appears that even though Antigua as an island may have reached the pinnacle of its sugar productivity in 1753 (Lowes 1994a), the Codrington estates continued to expand their production such that yields were higher in the 1760s than before. No major climate events are known for this decade, suggesting that a reliable climate that was never too dry or too wet created conditions favorable to a decade of successful sugarcane crops.

Multiple years of extreme weather conditions were recorded in the 1770s. The period of 1771–1774 was unusually wet, while 1776–1779 was a period of multiple droughts. A hurricane in 1772 was "one of the most destructive hurricanes in decades" (Ragatz 1928:134) for the Leeward Islands; buildings were flattened and crops were decimated. Correspondence from Betty's Hope to the Codrington family reported that a small sugarcane crop was expected in 1779 due to prolonged drought conditions (Lowe 1951). Sugarcane yields for this decade displayed much variation, possibly as a result of extreme weather. Yields were low in the first half of the decade, increased in 1774, then dropped precipitously in the second half of the decade. The year 1775 had normal precipitation in Antigua; thus it follows that the transition between extremely wet and dry periods was the most successful for crop production. The American War of Independence (1776–1783) further exacerbated the general decline of the British colonial sugar industry (Watts 1987:278–279). The implications of this conflict on the British West Indies were profound. Britain lost its command of the seas for the first time in the eighteenth century, as trade relations between the Caribbean colonies and American colonies on the mainland were disrupted, making supplies and food less available. Other consequences included increased operating costs, a decrease in sugar markets, and fewer sugar exports from the islands (Crist 1954:228; Ward 1978:209; Watts 1987:278–279). Reflecting this time of turmoil, the six-year moving averages of sugar yields for this decade show trends of decline, a contrast to the previous decades, when average yields increased or remained constant.

Sugar yields in the 1780s were also quite variable but were generally higher than those in the 1770s, perhaps indicating a recovery following a decade of climatic variation and political unrest. However, the 1780s were also a decade of extremely dry weather conditions. In 1780, Antigua experienced both a drought and a hurricane; a drought occurred in 1782–1783, and another occurred in 1788–1789. These years of dry conditions are associated with declines in crop yields. Records are not available for 1780 or 1781, but yields dropped in 1783, then increased in 1784 to 1787, then declined again in 1788. The declines in crop yields correspond with years of recorded drought, while the increases in the middle of the decade correspond with years of normal precipitation levels. The American War of Independence also ended in 1783, coinciding with an increased crop yield the following year. Betty's Hope also appears to have continued its expansion; in 1780, the plantation had 393 enslaved laborers (compared to 227 in 1751), indicating an increase in labor capacity (see chapter 10). This reflects an increased demand for labor investments into the plantation land and sugar production. However, it was reported in 1781 that the enslaved population was too low for the 1,100 acres of estate lands and for annual production of 800–1,000 hogsheads of sugar (Lowe 1951:26). This suggests that Betty's Hope needed a larger labor force to maintain production levels. The six-year moving averages of sugarcane yields for this decade reveal that the decline in average yields from the previous decade continued, then yields increased from 1785 to 1790, producing a net increase by the end of the decade.

Historical records are not available for the period of 1791–1801, but it is known that the drought from the end of the previous decade continued in 1790 and 1791. Correspondence from the Codrington estates in Antigua recounts poor crops in 1790 due to the drought. Correspondence also reports that rain in 1791 provided relief from the dry conditions and the expectations that good crop yields would follow. Without historical records of sugarcane crops, it is difficult to determine the changes to the Codrington estates during this decade.

Historical records return in 1802, reporting higher yields than the last recorded yield in 1790. However, yields dropped significantly in the beginning of the decade. The year 1803 marked the first year of the Napoleonic Wars (1803–1815) between France and a number of opposing coalitions, including Britain, which remained involved throughout the entirety of the war. This conflict may be related to declines in sugarcane crops in 1804 and 1805. During this conflict, European consumption of government-subsidized beet sugar increased, creating competition and reducing the market for Caribbean sugar (Crist 1954:228). In 1807, the British Parliament passed the Abolition of the Slave Trade Act, which abolished the slave trade in the British Empire but did not abolish slavery itself (Porter 1970). This act would have cut off the influx of new labor in the West Indies and may be related to declines in sugarcane yields in the late 1810s. While sugarcane yields for this decade were quite variable, the six-year moving averages of yields indicate an overall trend of slight decline in this decade. Although the Codrington estates did not appear to decline following Antigua's peak in the middle of the eighteenth century, it appears that crop yields begin to decline in the nineteenth century.

Historical records are absent in 1815 and 1816, but sugar yields continued to decline in the 1810s. With the end of the Napoleonic Wars in 1815, British sugar prices dropped as French colonial production recovered, Cuban harvests grew in size, and cheap East Indian sugarcane and European sugar beet emerged on British and colonial markets (Watts 1987:282). A hurricane was recorded in 1812, possibly contributing to declines in the first half of the decade. Historic yields declined from 1810 to 1813, coinciding with reports from Betty's Hope that poor weather caused a poor crop (Lowe 1951:59). The rise in yields a few years later coincides with a report of a good crop at Betty's Hope in 1817. The six-year moving averages from this decade indicate an overall decline in yields for the 1810s, despite annual variation.

Sugar yields continued to vary in the 1820s. A decline in the beginning of the decade coincides with a period of drought from 1820 to 1822. Correspondence to the Codrington family corroborates this decline, reporting that the 1821 sugar crop was small due to drought but that yields improved in 1823. It was also reported that Betty's Hope had 310 enslaved laborers in 1827. This figure is lower than that reported in 1780 (393 enslaved laborers). Along with the declines in crop yields, the reduced number of laborers reflects a decrease in production capacity at Betty's Hope. The smaller labor force could be a product of the ban on slave trade two decades prior, which would have cut off the supply of new laborers. During this time, plantation owner absenteeism soared to approximately 70% during the 1820s (Watts 1987:283), and the average estate profits fell to less than 6% (Watts 1987:283). Antiguan sugar production waned by the first few decades of the nineteenth century, and the broader island-wide trend seems to be reflected in the crop yields from the Codrington estates.

It is clear that Antigua, along with the rest of the British West Indian sugar islands, experienced decline from the profitable and wealthy period of the mid-eighteenth century, when Parliament passed the Slavery

Abolition Act in 1833, decreeing that freedom should be granted to enslaved Africans in British colonies on August 1 of the following year (Watts 1987:469). Interestingly, despite dire predictions to the contrary, Antigua remained prosperous after emancipation (Lowes 1994a:4). The average export quantity from 1829 to 1833 was 12,189 hogsheads, while from 1834 to 1838, it increased to an average of 13,545 hogsheads (Lowes 1994a:10). In the early 1840s, it was reported that despite droughts in the 1830s, sugarcane crops were larger than those before emancipation (Lowes 1994a:10). However, Antigua's annual sugar exports declined steadily in the nineteenth century, and the island faced a crisis in the 1890s, when a drastic decline in sugar production coupled with low prices on the world market triggered an economic decline from which the island never recovered (Lowes 1994a:4).

Geoarchaeological Assessment

Our geomorphological survey revealed the presence of four well-defined landforms (following nomenclature outlined in Waters 1992): T_3 (oldest and highest terraces, most eroded), T_2 (generally stable terraces with high agricultural potential), T_1 (lowest and youngest terraces, highly dynamic), and T_0 (active fluvial channels). Each landform is distinguished by a unique rate and degree of sediment additions and losses, which together can provide an overall representation of regional soil integrity. Figure 8.4 shows the distribution of these landforms relative to the historical division of plantation fields around Betty's Hope as recorded in the 1755 map from the Codrington Papers. As shown in the overlay, Betty's Hope and the contemporary town of Pares (located immediately to the west of Betty's Hope) occupy T_3 landforms, while T_0 and T_1 landforms follow the general path of Mercer's Creek as it winds around the west and north sides of the main settlement at Betty's Hope. Finally, T_2 terraces are broadly dispersed throughout the area and account for most of the locations of the 1755 fields. Figure 8.4 also shows the locations of our auger probes and the resulting soil organic matter levels determined through loss-on-ignition analysis.

The T_3 landforms reveal the greatest degree of degradation and erosion, often displaying a partial or complete loss of the B horizon. For example, augur probes 17, 18, and 19 all have thin (ca. 2 cm thick) and weakly developed A horizons overlying weathered bedrock (C) with no evidence

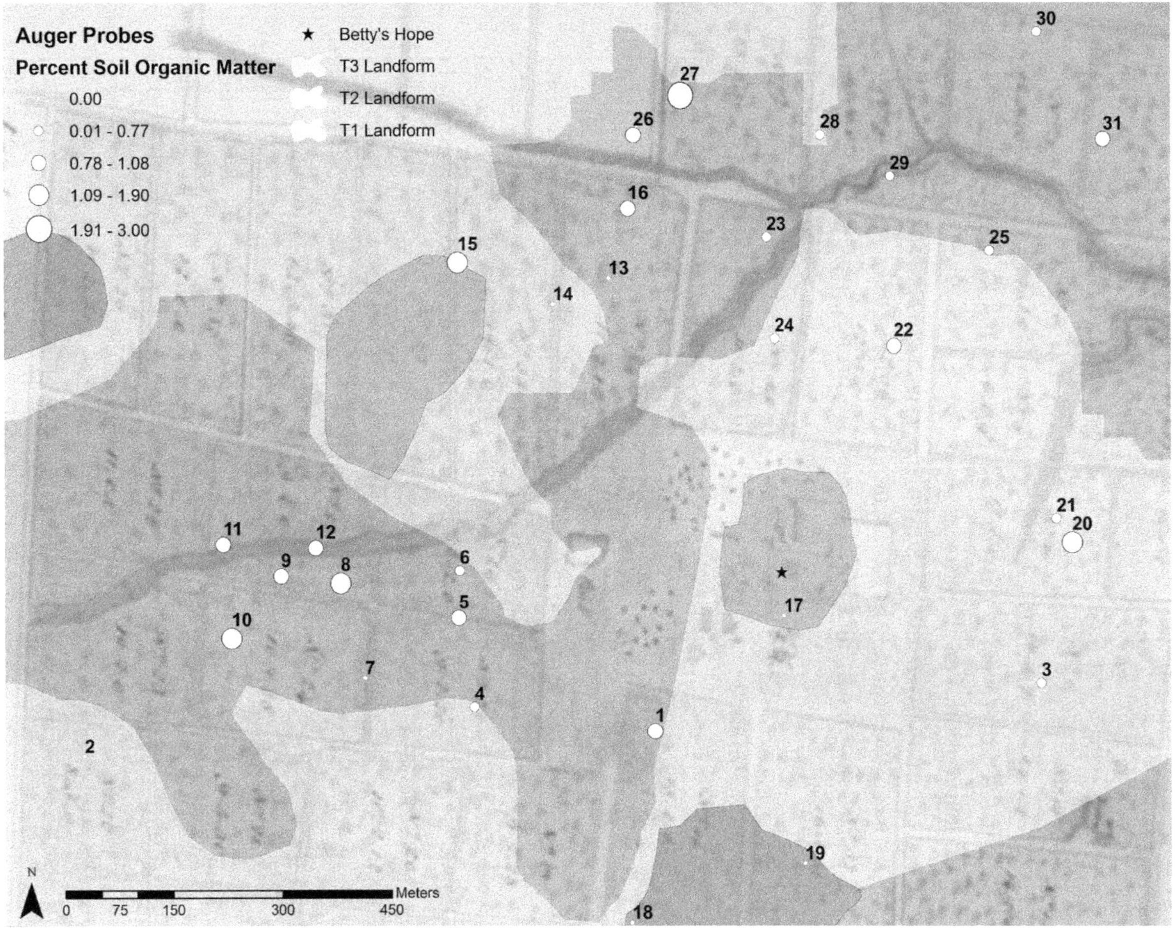

Figure 8.4. Geomorphological map of the area surrounding Betty's Hope, showing the distribution of landforms relative to the historical division of plantation fields as recorded in the 1755 map of the estate. The map also shows the locations of auger probes in the study and the corresponding levels of soil organic matter.

for subsoil. In all three cases, the A and C horizons are significantly depleted (one or more standard deviations below the mean) in transition metals (Ti, Mn, Fe, Zn). However, these areas are also enriched (one or more standard deviations above the mean) in potassium (K) and have organic matter ranging from 0.9% to 3.5% in the A horizon. This might be explained partly by current land use, which is predominantly residential. Overall, our geomorphological analysis of T_3 landforms reveals that they are the oldest and most stable in the region, and thus subject to greater cumulative rates of erosion over time.

The T_2 landforms typically have highly developed subsoil horizonation, such that weakly developed B horizons might indicate some level of degradation. For example, auger probes 2, 14, and 32 all show horizonation more typical of T_3 landforms with a sharp boundary transition between the A and C horizons along with depletion of transition metals (especially Ti, Mn, and Fe). Alternatively, probes 3, 4, 20, and 22 all have moderately developed subsoils and no evidence for depletion of metals and low soil organic matter, ranging from 1.0% to 2.0%. Probe 21 has a more developed argillic subsurface soil horizon (A–1Bt–2Btk–C) and a thick (ca. 10 cm) humus with 7.6% organic matter. In general, the T_2 landforms are more stable than T_3 landforms, but some areas are also subject to erosion caused by a lack of soil integrity, likely resulting from past and potentially current land use.

The T_1 landforms are generally the most dynamic and have been subjected to much more historic and contemporary anthropogenic disturbance. Generally, the T_1 landforms have deep A horizons and can contain thin, but complexly layered, B horizons, often with high clay content. For example, augur probe 8 revealed the presence of a deep (40 cm deep) A horizon and a shallow (10 cm deep) but stable B horizon, with significant enrichment in Ti, Mn, and Fe. Similarly, the A horizons in probes 6 and 11 were 20 cm and 27 cm deep, respectively. Probe 28 yielded a similar result but contained the presence of a B/C transition layer. Finally, probe 12 yielded a complex subsurface (1A–2At–1Bt–2Btk–C) as did probe 29 (A–1Bt–2Bt–3Bt–4Btk–5Bt). Organic matter ranges widely among all probes, from 0.8% to 3.1%. Many probes (especially 10, 11, 16, and 30) show significant depletion of K in contrast to T_3 landforms, which all showed enrichment.

To better understand the variations in soil properties among landforms and fields, we conducted a principal components analysis on seven plant

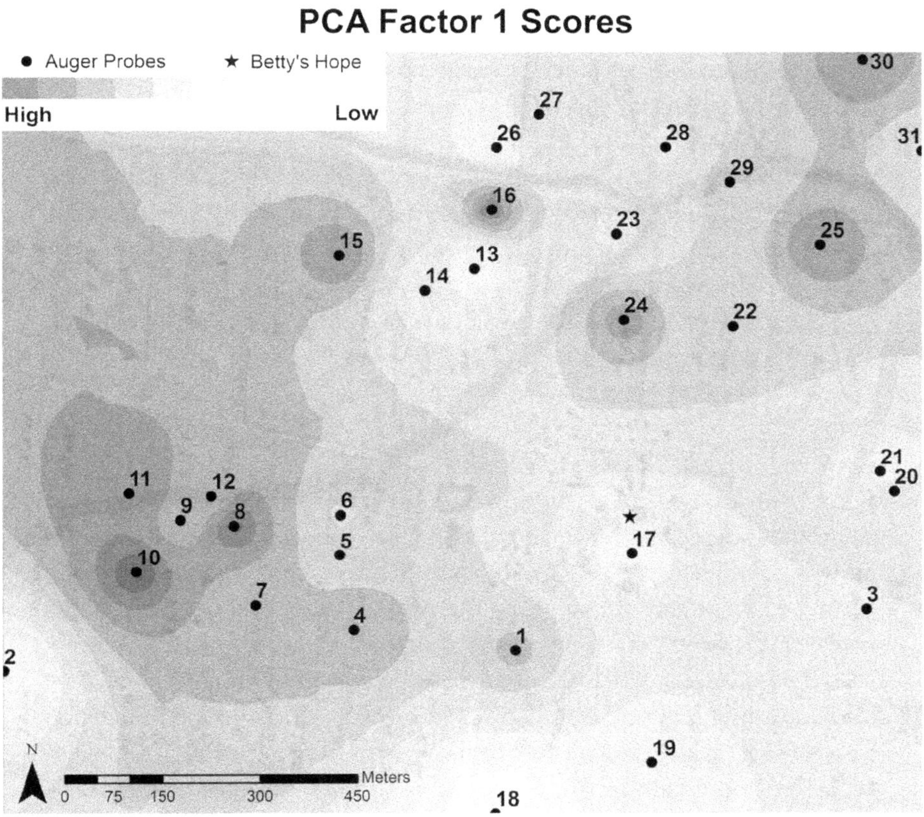

Figure 8.5. Spatial plot of interpolated (inverted distance weighting) factor scores from a principal components analysis with a covariance matrix. Darker hues correspond to higher concentrations of elemental signatures as identified by their component scores: Fe = .999, K = –.790, Ti = .820, Mn = .404, Sr = –.403, Ba = .208, Zn = .013.

macronutrients using a covariance matrix. The first component accounts for 99.55% of the variation (component scores: Fe = .999, K = –.790, Ti = .820, Mn = .404, Sr = –.403, Ba = .208, Zn = .013). Figure 8.5 is a spatial map of interpolated component scores based on inverse distance weighting, with darker hues corresponding to higher values of the combined elemental signatures. From these data, we can infer that the highest levels of soil degradation, which correspond to the lowest (light colored) levels of component scores on the map, are associated with Betty's Hope settlement in the center of the map and the ridge to the south connecting

Elliott's Estate. Soil degradation can also be inferred for the north and east of the town of Pares. The lowest levels of soil degradation, which correspond to higher (darker hues) levels of component scores, are associated with active agricultural fields to the south of Pares and to the northeast of Betty's Hope. In informal interviews with the owners of these fields, we learned that these areas have been farmed continuously at least since the 1940s and possibly earlier, suggesting that continued maintenance and use of these particular fields aided soil health. In contrast, fields with higher levels of degradation on the map have been subject to many different uses over the past century, including general abandonment, human settlement, and cattle pasturage.

Discussion

Both historical records of sugar yields and geoarchaeological data reveal a complex sugarcane legacy across the contemporary landscape. The examination of historical crop yields and the subsequent comparison of the simulated yield/acre provided by the EPIC model reveal that the EPIC model does not reflect accurately the changes in historical yields. The EPIC model predicts that yields should decline dramatically and ultimately stabilize at a much lower yield. However, recorded historical yields defied this pattern and continued to exhibit growth long after the takeover of sugarcane monoculture. The records show annual variation with a trend of yield increase from the beginning of available records in 1707 until the end of the eighteenth century and then a subsequent decline in yields after 1800. The deviation of historical yields from the simulated yields suggests that the Codrington plantations may have pushed the landscape to the limit to produce as much sugarcane as possible.

The historical yields also display variation associated with major political and climatic events. The effects of extreme wet or dry periods coincide with sharp increases and decreases in sugar yields. These effects are particularly acute in the 1770s and 1780s, when alternating years of too little and too much precipitation coupled with the disruption caused by the American War of Independence coincide with two decades of wildly variable sugarcane yields. Despite the increase in annual variation during this period, the six-year averages from these decades reveal that average sugar production was actually higher in these years than in the beginning of the century. Increased production coincided with an expansion of the sugar works at Betty's Hope in addition to increases in the enslaved population. It is clear from the historical records that sugarcane cultivation was beginning to destabilize; although more sugar was produced than in previous decades, crop yields were becoming unpredictable. In 1781, the middle of this period of dramatic crop variation, a letter from the supervisor of Betty's Hope to the Codrington family in England reported that the Codrington estates collectively had too few enslaved Africans to support sugar production. Although there were 393 enslaved Africans working at Betty's Hope at this time, production on the Codrington estates was too massive to be sustained by the available labor force. The amount of human effort put into the land to prepare the fields, plant the cane, tend to the cane, and harvest the cane, and the struggle to maintain consistently high levels of sugar production suggests increasing instability in the human-environmental system.

The Codrington estates, like the majority of Antiguan plantations, manured their fields and rotated them between periods of production and fallow in order to increase productivity and allow the soils a reprieve from consistent farming (Martin 1784; Museum of Antigua and Barbuda 2005; Sheridan 1960:133; Watts 1987:425). Continued investments into the landscape intended to improve sugarcane crop yields. However, Antiguan plantations over time drastically altered the landscape in order to make it ideal for sugarcane growth and subsequently dependent on human activity for its maintenance.

Despite efforts of Antiguan planters to stave off decreasing yields over time (Sheridan 1960), the sugar industry ultimately reached a point after which production levels could not be sustained (Watts 1987). Sugar was never again as productive or profitable as it was in the mid-1700s, though sugarcane cultivation persisted in Antigua until 1972. The long-term adherence to the sugar industry can be explained by the concept of path dependency. Path-dependent societies continue to follow a course of action based on tradition and practice—or short-term "least cost"—even if other alternatives are possible and potentially more desirable in the long term (Chase and Chase 2014; see also Fox 2014). The commitment to sugarcane monoculture in the early eighteenth century set Antigua on a course of pursuing great wealth at a high cost, as the island's economy subsisted on the production and export of sugar and its byproducts. Even after Antigua reached the pinnacle of sugarcane productivity, Antiguan plantations continued to produce sugar as annual yields progressively declined in the latter half of the eighteenth century, with sharper declines from the 1890s until the end of sugarcane cultivation in the early 1970s.

The adherence to sugarcane monoculture is visible in the sugar yields recorded in the Codrington Papers (NAAB Codrington Papers). Antiguan sugarcane peaked in the mid-1700s, but the Codrington plantations continued to produce sugar at higher quantities. According to the Codrington Papers, the year 1766 was the peak for sugar production. Additionally, the Codrington Papers indicate that yields in the second half of the eighteenth century were larger than those from the first half. This suggests that although the heyday of sugar production may have ended for Antigua, the Codringtons increased their production capacity by expanding sugar processing facilities, increasing the number of enslaved laborers, and producing larger annual quantities of sugar and rum (Goudge 2017; Wells et al. 2018a). In doing so, it appears that the Codrington estates created landscape conditions conducive to degradation and erosion such that the landscape came to rely on constant human maintenance. This is evident in the dramatic changes to sugar yields in the 1770s and 1780s, when climatic and political events created an unpredictable context for sugarcane cultivation.

The geoarchaeological data suggest which parcels of land experienced the greatest declines in soil productivity. Our survey and soil analysis indicates that the area of the Betty's Hope principal settlement and the fields to the southeast located on T_2 and T_3 landforms currently contain the most degraded soils, as indicated by topsoil loss (indicating landscape instability), low levels of organic matter, and nutrient leaching. These areas are currently not under cultivation and do not appear to have been involved in agricultural production during the past century. Alternatively, the fields to the southwest and northeast of Betty's Hope are currently under cultivation and, according to local farmers, have been in agricultural production since the nineteenth century. Notably, these fields are associated almost exclusively with T_2 and especially T_1 terraces that have developed soil horizons (indicating landscape stability), generally higher levels of organic matter, and enriched nutrient levels. These patterns support the historical data, which indicate that productivity could be sustained beyond that predicted by the EPIC model when the engineered soilscapes were carefully maintained. When such fields were not invested with the level of needed maintenance, such as the post-Emancipation period of the late 1800s, crop productivity fell dramatically. The last Antiguan sugar refinery closed and sugarcane production ceased in 1972, suggesting that recent erosional and depositional events documented in the region (Tricarico 2019; United Nations 2005; Weaver 1988; Wells et al. 2018b) may be related to the cessation of landscape maintenance. Pratt's (2015) analysis of soil catenas adjacent to Betty's Hope, which documents as much as 20 cm of soil accumulation on T_1 terraces, supports this observation.

Conclusions

The annual sugarcane yields recorded in the Codrington Papers indicate that historical crop yields were more dynamic than those predicted by the EPIC model. The deviation of historical yields from the simulated yields suggests that plantation agriculture may have been pushed to its limits. Consistent overuse of the landscape likely created ecological conditions that increased the fragility of soil health. As recent historical ecological research suggests (Fisher 2005; Håkansson and Widgren 2014; Wells et al. 2013), some highly anthropogenic landscapes may require constant maintenance in order to remain stable. Sugarcane agriculture, which requires large amounts of labor to plant, tend, and harvest, demanded an increasing amount of labor and investments in the Antiguan landscape as plantation owners struggled to extract as much sugar from their plantations as possible (see chapter 7). Chronic overuse of plantation lands for nearly three hundred years created a landscape dependent on human labor to remain stable enough for cultivation. The consolidation of Antiguan sugar estates in the mid-twentieth century represents the "abandonment" of the island's plantation sugar industry. The abrupt cessation of large-scale sugarcane cultivation—and the accompanying labor inputs into the landscape—may have precipitated the recent land degradation in Antigua. By 1972, the landscape was reliant on human maintenance, and the sudden and dramatic reduction to those investments caused rapid destabilization of a vulnerable landscape.

PART IV

Enslaved Life at Betty's Hope

9

"Choicest of the Cargoe"

Antigua, the Codringtons, and the Slave Trade, ca. 1672–1808

JAMES F. DATOR

In November 1718, the new heir of Betty's Hope, William Codrington I, sent directions to his attorneys in Barbados about his ship *Pearl:*

> You are on the arrival of my ship Pearl to pick 10 of the choicest of the Cargoe for my plantation in Barbados of Negroes & place them on my plantation. Then you are to pick out 20 of the next choicest of the whole Cargoe of Slaves 10 men & 10 women for my Plantation in Antigua to be sent to Antigua in my said ship. (Codrington Correspondence:C2)

In instructing his attorneys to purchase the "choicest" Africans from his ship, Codrington was participating in the booming separate slave trade, which allowed him the legal right to purchase captive Africans and sell them, duty free, after 1712 (Pettigrew 2013). This change helped make Britain the largest distributor of enslaved Africans in the eighteenth century. Between 1672 and 1808, British slave traders shipped more than three million Africans across the Atlantic—40% of the entire trade (Voyages 2017a).

William Codrington was a rare kind of slave trader. Not only did he possess ships like the *Pearl,* but he also possessed plantations in Antigua, Barbados, Barbuda, and St. Kitts in the eastern Caribbean. Owning a ship and multiple plantations empowered the owner of Betty's Hope to circumvent aspects of the marketplace. If needed, he could instruct masters like John Green to target certain regions of West Africa for the "choicest" Africans and deliver them directly to specific island plantations like Consett, Rooms, or Betty's Hope in Antigua. Codrington could also order captains to sell enslaved Africans elsewhere, such as South Carolina. Plantation owners like Codrington who also owned slave ships thus created a kind of early-modern vertical integration, wherein they controlled multiple parts of the production process (Radburn 2015).

A study of the Codrington family's engagement in the slave trade raises three important questions: What did the Atlantic slave trade to Antigua look like? How did these larger structures of the trade compare to those dynamics faced by William Codrington and his successors? And what can we learn about the African captives?

Over the past several decades, scholars have produced voluminous literature on the Atlantic slave trade (Curtin 1969; O'Malley 2014; Williams 1994). Far less research, however, has focused on the trade to Antigua. Perhaps no greater tool has been developed in the past two decades than the website Voyages: The Trans-Atlantic Slave Trade Database. The database allows scholars to draw conclusions about the timing and volume of the trade, connections to specific regions of West Africa, and the gender and age composition of Africans forced to disembark at Antigua. Relying on this data set, this essay reveals that the Atlantic slave trade to Antigua between 1672 and 1808 was far greater than other scholars have realized.

Codrington's letter to his attorneys also underscores that more attention needs to be focused on the family's regional slave trading activities. This granular approach illuminates how the slave trade affected Africans and Afro-Caribbean islanders who labored on plantations like Betty's Hope. In the case of the 1718 journey of the ship *Pearl,* Africans who survived the Middle Passage were first sold to Barbados and then to Antigua. The unknown captives who survived the dislocation and violence of the Middle Passage arrived in Barbados only to have themselves torn away from their shipmates and forced to labor for Codrington in Antigua.

This was not an uncommon practice for the Codringtons. A close examination of family letters demonstrates that Africans came to Antigua and Barbuda by multiple routes. In fact, Antigua may not have been the last stop for captives on ships like the *Pearl,* especially those who survived the first years of seasoning. An inter-island trade existed alongside the broader Atlantic slave trade to Antigua, one that connected Africans and Europeans in a violent web of forced removal. As Gregory O'Malley (2014) has argued, the inter-colonial slave trade sheds light on how Africans experienced dislocation outside of a Middle Passage paradigm. Codrington archives thus highlight how market dynamics and planter preferences for

Africans influenced the private trade. These documents also reveal the lived geographies that enslaved Africans and Caribbean-born slaves experienced when they were sold between colonies. Read together, these two dynamics show that both the violence of the regional and international slave trades shaped culture formation on the Codrington plantations.

The Africa-Antigua Slave Trade

Analyzing information from Voyages provides insight into the dynamics of African provenance, gender composition, and age differences among the people sold to Antigua. The Voyages database is not exhaustive, however. Some African cargoes and voyages will never be known. The ways the ship trajectories are classified in the database also makes analysis difficult. Africans bound for Antigua experienced multiple routes to their final captivity (O'Malley 2014). Indeed, at least 20 of the ships that are categorized as having Antigua as a "Principal Region of Slave Landing" between 1672 and 1808 stopped first at neighboring islands like St. Kitts, Barbados, and Dominica (Voyages 2017b).

Tracing the routes of Africans bound for Antigua is therefore not as simple as checking variables in the database and reading a table. Instead, the number of Africans sold at transshipment points must be considered when evaluating the data. Shipboard mortality rates also influence estimates. The Voyages database does not specify the gender composition nor age of the enslaved by point of disembarkation but rather by ship. Thus, the number of captives arriving at Antigua is difficult to conclude since their sale was not equal across different ports. Tracing the African origins of enslaved people sent to Antigua is also difficult because the record does not note provenance and point of disembarkation at the individual level. Slave traders, moreover, stopped at multiple ports in West Africa (Rediker 2007). The record also obscures how gender and age influenced plantation life during the era of the trade, which had a crucial bearing on African cultural patterns in Antigua. The historical record has thus left a fog of imprecision that continues to obscure historians' abilities to recover Antigua's African roots.

There are additional complications that veil Antigua's connection to Africa. For decades, historians and anthropologists have debated whether African ethnic categories such as Coromantee or Mina are useful for assessing the presence of African culture (Hall 2005; Law 2005; Morgan

1997; Ramos 1941). Some scholars argue that these ethnic signifiers are invented tropes stemming from Western travel literature, colonialism, and anthropology (Hartman 2006). Other scholars argue that Africans drew on their pasts to regenerate new ethnic affiliation in the Americas (Gomez 1998; Rucker 2015; Sweet 2013). Without a doubt, analyzing the African presence in Antigua through perceived ethnic boundaries has its shortfalls, especially because regional categories are not placeholders for group identity. Historians of Africans in the colonies need to study the Atlantic context, but also the internal social divisions, political tensions, and relational identities that existed within Africa if we are to paint a fuller picture of colonies like Antigua. Debates about whether ethnic signifiers and "cultural areas" are Western, creolized, or African will certainly continue. At a minimum, however, excavating African pathways and the gender and age composition of the enslaved provides an entryway into African ethnogenesis in Antigua (Gaspar 1985; Gomez 1998; Mintz and Price 1992).

One major finding of this study stands out. The extent of African imports to Antigua is far greater than David Barry Gaspar's estimates, which suggest the total number of Africans sold to Antigua between 1671 and 1763 hovered around 60,820 (Gaspar 1985). Data analyzed for this study point to a number 39% larger. At least 84,516 Africans are now known to have disembarked at Antigua during this period (Voyages 2017b).

Extending our analysis to 1808 broadens the picture. Between 1672 and 1808, at least 648 slave ships delivered Africans to Antigua. This number includes vessels that stopped at Antigua first and ships that stopped elsewhere beforehand, but it does not include six ships seized by the Royal Navy or the untold number of illicit vessels that reached Antigua after 1808. Close inspection of the 648 known slaving voyages indicates that slave traders embarked at least 156,348 Africans on ships before they set sail toward Antigua. An untold number were sold or died before reaching Antigua. Of total embarked, roughly 126,812 disembarked. At least 3,000 people on these ships died during the Middle Passage (Voyages 2017b). What happened to the other 26,536 embarked captives is lost to the violence of archival erasure.

Africans in Antigua came from every major slave-trading region. Of the 648 known voyages, records reveal that about 22% of Africans came from the Bight of Biafra. The region spans nearly 1,000 km of coastline and encompasses modern-day Nigeria, Cameroon, Equatorial Guinea,

Table 9.1. Africans Embarked on Ships to Antigua by Broad African Region, 1672–1808

Principal Region of Embarkation	Ship Count (n)	Number of Africans Embarked	Percent of Total
Bight of Benin	27	7,963	5.09%
Bight of Biafra / Gulf of Guinea Islands	111	33,835	21.64%
Gold Coast	82	20,233	12.94%
Senegambia / Offshore Atlantic	51	6,865	4.39%
Sierra Leone	29	5,550	3.55%
West Central Africa / St. Helena	34	10,849	6.94%
Windward Coast	64	14,129	9.04%
Other Africa / Unknown	250	56,924	37.94%
Totals	*648*	*156,348*	*100%*

Source: Voyages Database.

and Gabon. The second-largest group came from the Gold Coast, which stretches about 600 km along the coast from modern-day Ghana to Benin. Slave traders along the Windward Coast also sent more than 10,000 people to Antigua. After 1672, ports along the Bight of Biafra, the Gold Coast, and the Windward Coast delivered nearly 47% of Africans sold to Antigua. The Rice Coast, the Bight of Benin, and West Central Africa accounted for nearly 20% of the trade after 1672. Sources for 250 voyages, however, do not specify African provenance. This gap is represented in the row listed as "Other Africa / Unknown" in table 9.1 (Voyages 2017b).

This longitudinal perspective glosses over the rhythms of the trade, which influenced Antigua's changing cultural composition. Enslaved Africans from these regions arrived in Antigua at different rates; some regions dominated the trade for years and then declined. Ship records, however, make these rhythms difficult to uncover. Of the 648 known transatlantic voyages that touched at Antigua, 11 stopped at three African coastal ports to load enslaved people, but only one of the ships' records specifies the numbers of captives loaded from each port. Another 46 ship captains anchored their vessels at two different African ports, and of these, only six kept records by port. Estimates are also complicated because traders did not always account for African deaths by ethnicity.

Year-to-year analyses still yield fruitful information, however, about West Africans in Antigua. Africans from the Gold Coast arrived in Antigua in greater numbers than any other region between 1672 and 1740

Table 9.2. Africans Disembarked at Antigua by a 60-Year Period and Region

African Region	Africans Disembarked 1672–1740	Africans Disembarked 1741–1808	Percent Change
Bight of Benin	5,068	1,570	-69.02%
Bight of Biafra / Gulf of Guinea Islands	3,783	22,578	496.83%
Gold Coast	7,522	7,948	5.66%
Senegambia / Offshore Atlantic	669	4,971	643.05%
Sierra Leone	784	3,840	389.80%
West Central Africa / St. Helena	1,637	7,371	350.27%
Windward Coast	0	11,429	
Other Africa / Unknown	28,693	18,949	-33.96%
Total Disembarked (% Change)	*48,156*	*78,656*	*(63.34%)*

Source: Voyages Database.

(7,522; see table 9.2). Nearly the same number of Africans from the Gold Coast arrived in Antigua between 1741 and 1808 (7,948). In contrast, no slave ships carried Africans from the Windward Coast directly or indirectly to Antigua prior to 1740. Yet the trade from the Windward Coast increased dramatically after the War of Jenkins' Ear (1739–1748). Traders sold an estimated 11,429 people from the Windward Coast to Antigua after the war, making it the second largest exporter of captives to Antigua behind the Bight of Biafra (22,578). Africans from the Bight of Benin disembarked at Antigua in greater numbers (5,068) during the 68 years prior to 1740, but far fewer thereafter (1,570). The decrease in trade from Benin stands out compared to other regions. Dramatic increases in captive exports after 1740 also occurred from the Windward Coast, Senegambia (643.05%), Sierra Leone (389.80%), and West Central Africa (350.27%).

These changes related to the growing demand for enslaved labor in the Caribbean, but they also underscore the shifting power of African states on the trade (Kea 2012; Law 2004; Thornton 1998). British access to captives at Whydah on the Bight of Benin, for example, decreased after the Kingdom of Dahomey seized the port in 1727 (Law 1986; Voyages 2017b). William Snelgrave, captain of the slave ship *Katherine Galley*, learned about the drama unfolding at Whydah and chose to benefit from the war (Snelgrave 1734). The same year that Dahomey captured Whydah,

Snelgrave purchased 606 captives from the nearby port of Jacquin. Fifty-six Africans died on the voyage, but Snelgrave still sold 545 people to Antigua (Voyages 2017b).

Antigua's traders maintained a vague idea of these shifting patterns because planters believed African stereotypes corresponded to vitality and work ability (Gaspar 1985; Morgan 1997). Antigua's planters thought that Africans from the Gold Coast and Whydah were hardier than Africans from the Bight of Biafra and Angola. In 1750, Antiguans even wrote to the Board of Trade complaining that they were having a harder time accessing Africans from these preferred locations, noting that prices for Africans from Whydah and the Gold Coast were high. The petitioners also argued that slave traders from Bristol and Liverpool preferred to trade in cheaper slaves from Calabar and Angola (November 8, 1750, NACO Antigua Sessional Papers:9/20). Antigua's planters even claimed that they would be "ruined" if they could not increase access to the Gold Coast and Whydah markets (November 8, 1750, NACO Antigua Sessional Papers: 9/20). Planter Samuel Martin was so desperate for Gold Coast Africans that he considered purchasing his own slave ship (Gaspar 1985).

Although their appeals were grounded in stereotypes, planter anxieties nonetheless reflected changes in the supply of Africans from these prized regions. As noted previously, Africans from the Gold Coast and Whydah were arriving in smaller numbers by 1750—the former in relative terms, the latter in absolute terms. What Antigua's planters failed to recognize in 1750, however, was that the changing political alliances on the African coast were shaping supply. Much as William Codrington himself would realize in the 1720s, accessing the "choicest of the cargoe" was easier said than done.

Finally, an analysis of age and gender patterns in the African trade to Antigua between 1672 and 1808 reveals critical information about the colony's captive laborers. Of the known voyages examined for this chapter, only 43 identified the captives as adult men or women; 37 voyages also identified the number of captive boys and girls. Adult men outnumbered all groups. Men averaged 48.9% of the Africans on known ships. Adult women made up about 31% of all Africans traded to Antigua. Nearly 24% were children. Enslaved boys made up about 15% of all captives while enslaved girls comprised just under 9%. It is thus possible that nearly a quarter of all Africans who arrived at Antigua were children under the age of 14 years old (Voyages 2017b). The age imbalance on a few ships is striking. Snelgrave's *Katherine Galley* from Whydah sold roughly 545 people to Antigua in 1727, 83% of whom were children (Voyages 2017b).

The Codringtons and the Slave Trade

These broad patterns of the African slave trade to Antigua capture only part of the picture. Focusing on the Codringtons' activities illuminates a more specific story. Records of the Codrington family's role in the slave trade are extant after 1700, following the death of Christopher Codrington III and the takeover of Betty's Hope and other Caribbean plantations by William I. Like his relatives, William relied on the Royal African Company (RAC) to purchase Africans early in his slaveholding career. On July 24, 1702, RAC officials noted that William I purchased four girls from the ship *William & Jane* for £122 (August 23, 1708, NACO Board of Trade:152/7). These four unnamed girls were sold at Whydah (Voyages 2017b). A year later, William Codrington purchased another five unnamed men and five women from the RAC ship *Bridgewater,* which arrived at Antigua via Accra (August 23, 1708, NACO Board of Trade:152/7).

Warfare was another early way the Codringtons increased the size of their labor force in the late seventeenth and early eighteenth centuries. Both Christopher Codrington II and his son led expeditions against French territories in the 1690s and early 1700s. They captured people of color who were then forced to slave at Antigua. The number of captives the Codringtons took from French Saint Kitts and Guadeloupe likely extended into the hundreds and may have accelerated Antigua's sugar boom (Dator 2015). Whether William Codrington gained from these raids is unclear. An inventory for Betty's Hope dated May 1, 1715, suggests that he ended up owning captives from the French islands. One man is listed as "Peter Guadeloup," while a woman named "Old Betty" appears to have come from French Marie-Galante, both islands raided by his uncle more than a decade earlier (Codrington Corresondence:C2). Some of these names are listed in table 9.3, which provides a window into the African diversity at Betty's Hope in 1715.

Captives taken from nearby islands lived alongside Africans dislocated by the Codringtons on their own slave ships, which also sent rum to Africa. After the Peace of Utrecht and the end of the RAC monopoly in 1713, William Codrington invested heavily in slave trading. Rum production

Table 9.3. A Selection of Africans Who Labored at Betty's Hope, May 1, 1715

Women	Men	Girls	Boys
Old Betty Marigalant	Quacoe Barbados	Mary Barbuda	Peroe Creole
Mary Barbuda	Collomante Quaw	Accobah	Quaw
Ebbo Nan	Peter Guadeloup	Ann	Dick
Fibah Creole	Yamenna	Quashebah	Quashe
Betty's St. Kitts	Eboe Samson	Rose	Kitt
Pappaw Mary	Jackoe Creole	Mimbah	Cudgoe

Source: BLCP: D1610 C-2.

was central to his plan. He used rum sales to fund his parish taxes, to pay servant wages, and to acquire enslaved Africans. Most importantly, Codrington ordered his attorneys to use rum to purchase his own slave ships. In 1719, William I instructed his Antigua attorney to use rum to "buy a very good Sloop not less than 113 foot by the Keal or a good large Brigateen of snow." He wanted the ship to be "fitted out for the Gold Coast of Guinney." Load it, he ordered, as "deep as can swim with rum" (December 6, 1719, Codrington Correspondence:C2).

The Codringtons owned at least three slave ships in the early eighteenth century: the *Codrington*, the *Pearl*, and a ship named *Betty's Hope*. Several ships in the Voyages database shared the name *Codrington*, but only two were clearly owned by the family. One ship, a scow called *Codrington*, was owned by William. This ship arrived at Antigua on November 20, 1727. It disembarked 139 people from "Africa" (NACO Board of Trade:152/16). A second *Codrington* slaver came from the Gold Coast and Senegambia and was captained by a "Green," likely John Green. Records for the second ship provide only vague details, which suggests either a discrepancy in the government account or a ship owner who really liked naming ships after his family (Voyages 2017b). By 1728, another ship called *Codrington* had a different captain (unlike the 1727 vessel from Bristol, this ship was registered in Rhode Island). John Williany captained this ship and disembarked 219 Africans (origins unknown) at Barbados for Codrington. Williany does not appear to have landed at Antigua (Voyages 2017b).

The Codrington family benefited from at least nine distinct purchases from West Africa between 1713 and 1735. More than 1,900 captives were forced onto Codrington-owned ships during this 20-year period. It is likely that William Codrington personally ordered each journey. While he was in England in 1715, he instructed attorneys for his Betty's Hope and

Cotton plantations to acquire at least 60 Africans from Captain Green. "When please God Capt. Green arrives," he wrote, "put 25 of the choicest men on Betty's Hope." He also ordered his assistants to obtain 5 women, 10 boys, and 10 girls from Green. Codrington also wanted 5 boys and 5 girls placed at a third plantation, "the Cottin New Work." Africans over 15 were to be "disposed of" for cash or sugar. His attorneys were forbidden from taking bills of exchange, the eighteenth-century equivalent of a check ("Instructions," Codrington Correspondence:C2).

Profit, not patriarchal duty, drove William I's decisions as a slave trader. In the 1730s, Codrington began directing his ship *Betty's Hope* to South Carolina, a new, booming market. William Codrington consciously redirected his slavers toward the rapidly growing colony in 1738, just before his death (Voyages 2017b). This moment was when direct trade from Africa to the northern colonies began (Wood 1975). Seeking profit, Carolina's rice plantations seemed like a good option for the absentee slave trader.

William I also had a proclivity for purchasing African children, like his fellow planters at Antigua. Nearly 25% of enslaved Africans who arrived at Antigua were children. Several years after William's death, attorney Benjamin King advised William's son—who lived most of his life in Europe—that enslaved children were the best Africans to purchase for Betty's Hope. Believing that Betty's Hope needed at least 60 more hands in 1740, he encouraged the heir to obtain children from Africa. "I would not Recommend purchasing the full grown," King wrote from Antigua, for "such Slaves are allways Obstinate and Stubborn & Seldom comes on kindly to Labour. . . . Negroes from 12 to 15 Years of Age are best," he explained, "and are most Reasonable" ("Representation of the Condition," Codrington Correspondence:C5). The life and experience of enslaved children in Antigua thus demands further study.

The Codrington family's direct engagement in the Atlantic slave trade appears to have ceased after William Codrington's death. No references to Codrington-owned ships appear in records after 1739. It is possible that members of the Codrington family held investments in insurance companies and shipping firms. If so, these documents do not appear in the Codrington papers consulted for this chapter.

The family's withdrawal from the vertical-integration production model does not mean they stopped participating in the slave trade. In fact, William's heirs followed in his footsteps. As William I invested in slavers like the *Pearl, Codrington,* and *Betty's Hope* and instructed his attorneys to purchase Africans, he was simultaneously displacing his captive workers. Codrington ordered these same attorneys to redistribute his captives across his plantations, to sell them or rent them to other planters, and to deliver them as gifts to his family members at other colonies. His heirs followed suit.

Age was the primary reason the Codringtons moved captives from one plantation or colony to another. In 1717, William I instructed his attorneys to remove one woman from Betty's Hope and to send her to Barbados. Codrington wanted to put "Old Betty," known as "Watty's wife," on the first vessel to Barbados, where she was to live the rest of her life on his mother's plantation (March 1, 1717, Codrington Correspondence:C2). He also asked his attorneys to move Africans to a plantation purchased from Thomas Warner "to plant ginger and make a ginger Plantation of it forthwith." These unnamed Africans were not supposed to be the best in the field, but rather the "refuse" from other properties (March 1, 1717, Codrington Correspondence:C2).

Codrington thus shaped his family wealth through the forced dispersal of African captives. These decisions were motivated by the very same instincts that shaped his role in the slave trade proper. Like the transatlantic trade, moreover, his decisions separated African families. Although it is possible that separated kin in Antigua could still reunite on Sundays or market days, inter-island transfer severed family and social bonds. These experiences created a kind of second Middle Passage. Slave narratives like Olaudah Equiano's *Interesting Life* speak directly to the violence this forced dislocation caused to the enslaved, and thus beckon further research (Equiano 2003).

Following William Codrington's death in 1738, Betty's Hope became the property of his son, William II, who lived until 1792. Available records do not reveal what happened to the slave ships his father owned. William II left plantation management and the slave trade to his attorney Benjamin King and his manager Samuel Redhead. Both men worked to sway the absentee's decisions about purchases and the distribution of labor. Naïve, William II authorized sales but rarely directed them. After receiving an analysis of his plantations indicating that Betty's Hope was in "absolute need of 60 Negroes," he questioned whether slaves from Barbuda should be transferred to the Antigua plantations (1740, Codrington Correspondence:C5). Codrington pushed back against his attorney's recommendation, but King purchased "five New Negroes" anyway on October 8, 1744, charging them to Codrington's accounts for £168 ("Accounts," Codrington Correspondence:A2). It is not clear where these five Africans were from, but the Voyages database details that at least seven slavers anchored at Antigua in 1744—two from unspecified ports, two from West Central Africa, two from the Bight of Biafra, and one from Sierra Leone (Voyages 2017b).

Distant and unlearned in sugar production, William II relied on advice about how to make up for the high death rate on his plantations. He considered purchasing, renting, or importing labor from Barbuda, usually pushing for the latter despite advice against it. In the 1750s, William II solicited advice from the famed Antiguan planter Samuel Martin, author of the manual *Essays Upon Plantership* (Martin 1750). Noting that Codrington's properties were in desperate need of "season'd Negroes," Martin advised him to rent rather than purchase Africans who had learned the basics of sugar and survived their first years of slavery. "I know from experience that it is a dearer method by half than putting on season'd Negroes," Martin related (June 4, 1752, Codrington Correspondence:C6).

Codrington rejected Martin's advice and chose instead to question his manager why he could not stock his Antigua plantations with captives from Barbuda, which did not grow sugar. Redhead pushed back by enlisting Joshua Crump, another planter from Antigua. Crump wrote to Codrington and argued that bringing captives from Barbuda to Antigua was a bad idea because "their labour on this Island differs so much from that of Barbuda, and by that means they become and troublesome and expensive to you." Crump advised Codrington to "take a chance" and purchase both "New Negroes" and "seasoned Negroes" to replenish his plantations (July 15, 1753, Codrington Correspondence:C7). Codrington agreed with Crump's theory, but he did not follow his advice.

The choice reflected the willingness of planters like Codrington to fuel the slave trade through purchases rather than improved conditions, which showed just how little they cared about Africans as people. It was precisely these kinds of decisions that animated local myths in Antigua about how the Codringtons used Barbuda as a "slave breeding" ground (Lowenthal and Clarke 1977). While historians have dismissed this claim, it is important to note that William II, like other members of his family, wanted captives from Barbuda to replace Africans who died in Antigua.

Codrington expressed deep skepticism about his managers' reasoning why captives from Barbuda could not be transplanted to Antigua for decades. Writing to manager Richard Oliver on March 8, 1780, Codrington described his "hopes that it might have produced Stock of Negroes to have assisted my Estates at Antigua" (Codrington 1779–1782). It is not surprising that Antiguans came to believe that the Codringtons used Barbuda specifically for providing black labor to Antigua. Though not a "breeding ground," the neighboring island did serve as a source for enslaved workers at the Codrington family's other plantations (see chapter 4). In the colloquial sense, there is thus truth to the myth that the reproduction of black labor at Barbuda fueled sugar production at Antigua.

Conclusion

Close analysis of the Codrington slave trade reveals how the family profited from the direct trade in Africans as well as the regional slave trade, especially when William I ran Betty's Hope. Yet details about African culture on Betty's Hope remain mysterious. While the study of archival material can yield new insights into the lives of eighteenth-century Africans and colonists in Antigua, archaeological evidence can reveal evidence from material culture that illuminates how and when Africans from so many different backgrounds generated new aesthetics and rituals. Conversely, an analysis of slave-trade data and letters by private slave traders can refine the archaeological context. Can the material culture found on plantations like Betty's Hope be understood in an Atlantic context and perhaps reveal certain temporal connections to specific places in West Africa? Does evidence of Afro-Antiguan pottery or beadwork show aesthetic changes that may reflect specific African influences (see chapters 10 and 11)? Unfortunately, both the slave trade and slavery in Antigua were defined by a dramatic elision of African diversity from the written record, leaving scholars of both disciplines with a fragmented record that obscures as much as it reveals. Despite these challenges, more multidisciplinary, collaborative work can unearth fruitful insights into this buried African past.

10

Enslaved Life at Betty's Hope

GEORGIA L. FOX

In the sociopolitical landscape of the twenty-first century, long after abolition and emancipation from plantation slavery, matters of racism and genocide remain unresolved. In that vein, the contributions of historical archaeologists to the dialogue are critical. They can offer substantive contributions to understanding the root causes of racial and ethnic conflict and provide thoughtful and insightful discourse to redress the past (Bates et al. 2016). This chapter seeks to provide as full a study as possible of the enslaved population at Betty's Hope through the archival and archaeological records. It is hoped that this chapter can serve as a basis for comparison, but also acknowledge that Betty's Hope had its own unique set of circumstances.

Property, Not People

The choice to resort to enslaved labor by European powers in the Caribbean was fateful, with terrible consequences for all involved. In a choreography of fear and threat, the Caribbean plantation operated as a human factory, where efficiency was paramount to a system vulnerable to natural and human-induced disasters. As owners of one of the largest agro-industrial systems on Antigua and Barbuda, the Codrington family exercised their power by flagrantly purchasing slaves in defiance of the Royal African Company's monopoly, the main conduit for Antigua (Dyde 2000:37, 39; see chapters 4 and 9).

As the floodgates opened to the transatlantic slave trade, over 12.5 million enslaved Africans were brought to the New World, with thousands brought to Antigua (see chapter 9; Deerr 1950: 279; Dunn 1972:140; Eltis

2001, 2006; Klein 2010:31–33; Pettigrew 2007:5–6; Tombs 2015:339; Voyages 2013). Antigua's initial enslaved population probably comprised Fanti and Asante groups, followed later by other groups, including Creole and Antiguan-born (Dyde 2000:31). One of the most compelling aspects of studying slave life on a Caribbean sugar plantation is reckoning with people as property, especially when listed with domestic animals.

As large estates like Betty's Hope consolidated their landholdings, importation of enslaved Africans increased to accommodate these changes. The first enslaved population at Betty's Hope probably dates to at least the Joan Hall period (see chapter 2). During 1700, when Christopher Codrington III was governor of Antigua, there were close to 10,000 enslaved Africans on the island and a population of about 3,000 whites (Dyde 2000:41).

Between 1720 and 1755, more Africans arrived, ranging to well over 50,000, with another wave occurring from 1755 to the 1770s, comprising another 30,000 to 40,000 enslaved people (Dyde 2000:41, 65–66). In concert with these numbers, importation of enslaved Africans to Betty's Hope and other Codrington estates increased when expansion efforts intensified. Table 10.1 indicates the number of enslaved Africans at Betty's Hope by year.

As property, enslaved persons were given Anglicized names, classified into distinct occupational categories, and assigned a monetary value. Names included Mimba, Nanno, Charles, Mary, Sarah, and others (Gaspar 1985:131; Handler and Jacoby 1996). Job duties were categorized as field laborers, domestics, head people, tradesmen, and those "employed in wharves, shipping, etc." (Higman 2000:370–371). Values for enslaved persons at the Codrington estates, including Betty's Hope, are shown in table 10.2.

At Betty's Hope, most field hands were predominantly women, about three-quarters of the female enslaved population, with the younger and less vigorous assigned to tasks suited to their stamina and abilities (Gaspar 1993:113). An account dated to May 15, 1740, reports 312 enslaved at Betty's Hope in the following categories: 63 women field hands, 37 women in miscellaneous duties, 27 tradesmen, 44 able field men, 33 large boys, 32 small boys, 23 large girls, 30 small males and infants, and 23 men "superannuated" (NAAB Codrington Papers:E5).

Other tasks required skilled sugar factory workers, coopers, blacksmiths, and livestock tenders and monitors for Betty's Hope water supplies

Table 10.1. Number of Enslaved Africans by Year at Betty's Hope

Year	Number of Slaves
1715	322
1740	312
1774	372/398
1779	370
1780	392/3
1790	382
1793	310
1806	350
1807	348
1809	350
1810	348
1811	340
1813	107
1814	342
1816	314/140
1817	311
1818	310
1819	308
1820	301
1826	312
1827	310
1830	289
1834	296[a]

Note: a. August 1, Day of Emancipation.

Table 10.2. Value of Slaves on Codrington Antiguan Estates, Including Betty's Hope

Year	Occupation	Value, per person (In £)
1774	Tradesmen	90£ ea.
	Field men and women	80£ ea.
	Boys and girls	60£ ea.
	Small boys and girls	35£ ea.
	Men and women of little service	20£ ea.
	Infant boys and girls	16£ ea.
	Women and girls in different employments	50£ ea.
1781	Adults	50£ ea.
	Children	30£ ea.
	Old *"superannuated"*	10£ ea.

(NAAB Codrington Papers:C2). Domestic staff served the Codrington family or managers, which included cooks, butlers and maids, housekeepers, footmen, nurses, seamstresses, washerwomen, and coachmen (see Lanaghan 1991[1844]).

Carpenters and stonemasons were important to Betty's Hope for the construction and repair of buildings, walkways, and walls; stonemasons are mentioned in the Codrington accounts, with an entry for 1781 stating that "Negro masons working on 5 new stills" (NAAB Codrington Papers:A5). Other occupations included saddlers, fishermen, day-watchers, and firemen (NAAB Codrington Papers:E6). Regardless of occupation, all duties required long hours to curtail any potential "trouble," with some time off for holidays (Gaspar 1985:131). In some instances, paid services at Betty's Hope were rendered by contracted free workers, such as a wheelwright (Rob Copeland, £70), bookkeeper (John Drew, £100), groom (Rob Gay, £50), and overseer (William Carrington £100), as noted in an account for 1780 (NAAB Codrington Papers:A8).

As with most plantations, hierarchies existed among the enslaved community based on the overall "occupational structure" of the plantation (Roberts 2013:275). At Betty's Hope, distinctions were made by position and skin color. Lighter-skinned individuals were placed in the domestic sphere and had greater access to better clothing, food, and housing, unlike field labor. This is partially supported by the servant's quarters in the great house compound.

The dehumanizing aspect of "fitness" was another form of property categorization at Betty's Hope and elsewhere. Higman (2000:369) estimates that 81.7% of enslaved Africans comprised active labor, except children under six years of age, and the old, weak, and infirm, who were designated as "ineffective" or inferior. In a letter to William Codrington II, manager Richard Oliver reported that Betty's Hope had "370 Negroes very mixed in quality rather inferior to the average of plantation negroes" (NAAB Codrington Papers:C12). As Antiguan Samuel Smith recounts (Smith and Smith 1986:28), "white massa didn't have much use for old worn-out slaves. . . . It was the custom of the Old Road slave massas to take the weak and sick slaves out to sea and throw them overboard. They didn't have any time to dig holes and bury them."

Working Conditions at Betty's Hope

Working conditions at Betty's Hope were harsh. Women field hands were divided into gangs of 40 to 50 to carry out the planting, weeding, fertilizing, and harvesting of the cane with only the simplest of tools (Dyde 2000:67). The adage at Betty's Hope was "One negro for every acre you cultivate" (NAAB Codrington Papers:C20/4, March 1790). Days were gruelingly long, with a short break midmorning and about an hour's rest at midday. Backbreaking, degrading, and physically exhausting, it was also humiliating, with the threat of the whip. Additionally, cane harbored vermin, and with its sharp edges, could generate cuts that became infected.

Harvest, or "crop time," required intensive round-the-clock labor (Davy 1971 [1854]:121). To avoid spoilage, extracting the juice from cane stalks quickly was critical. Crushing the cane in the mill was not for the faint of heart; it was dangerous work that required skilled hands and calm minds. The powerful jaws of the iron rollers might mangle hands and arms, while the rapidly spinning sails could sever human heads (NAAB Codrington Papers:C20/4).

In the boiling room, stirring and monitoring kettles of molten sugar could result in third-degree burns. Packing the sugar into large hogsheads and loading them onto wagons required strong backs and alert drivers to haul cargo over gutted dirt roads. Rum distillation had its hazards, too. The large distillation apparatus could explode, and distillers probably suffered from lead poisoning due to repeated exposure, as archaeological excavations there yielded copious amounts of lead piping (see chapter 12).

Bare Essentials

Clothing

As with most things for the enslaved at Betty's Hope, clothing and food were basic. Archaeologically, it is rare to recover textiles, so information about what Afro-Antiguans wore is left up to the documentary record and published literature. The Amelioration Act of 1798 established set rations for clothing allotment, which included a woolen cloth jacket and a pair of osnaburg trousers for men and a woolen shawl and an osnaburg petticoat for women twice a year (Goveia 1965:193). That said, this specific ration was not always adhered to, and clothing rations could vary from plantation to plantation (Higman 1984:223). Osnaburg (or osnaburgh) is a fabric made from the coarse linen family of flax, unbleached linen, or hemp, originally imported from the German city of Osnabrück. After 1730, this plain fabric was mainly produced by the Scottish for the export market (Durie 1993:71). The Codrington correspondence and accounts are filled with purchases of osnaburg, including "4 bolts of osnaburg," or "1 bolt of Oznaburg, 132 yards for the Negroes in the boiling and still house for aprons" (NAAB Codrington Papers:C28, A12). In one account for 1781, clothing for the enslaved amounted to an annual cost of £40 (NAAB Codrington Papers:A8).

Diet and Health

To live the life of an enslaved person on a sugar plantation meant survival in the strictest sense. Infant mortality, hardship on women's bodies resulting from backbreaking labor, punishment and violence, epidemics, poor nutrition, and calamities resulting from natural disasters all took their toll on the enslaved population at Betty's Hope and on the island. One of the more recent aspects of studying the lives of the enslaved is through diet and health. With the exception of human skeletal remains, which are not always available for study, hard evidence from which to reconstruct these aspects of enslaved life is often lacking, but such reconstruction is not impossible. Important clues do exist; it is a matter of teasing them out.

As traumatizing as the transatlantic crossing was for enslaved Africans, the fate awaiting them—if they survived the journey—was not much better. Kiple and Kiple (1980:199–200) suggest that by virtue of a low protein diet, enslaved Africans were already suffering from malnutrition upon their arrival to the New World. Once they arrived, conditions were not much improved. Fed on a daily ration of beans, Guinea corn and yams, supplemented with plantains, salt fish, molasses, and salt, the slave diet could be tedious (Higman 1984:206). At Betty's Hope, imported foodstuffs and rations shipped from the Codrington provision grounds on Barbuda made up a significant portion of the slave diet (see chapters 4, 5, and 6). An account for 1781 at Betty's Hope indicates the "maintenance of £100 slaves in food at £5 per annum is £500, out of which deduct 1/3 for provisions raised on the estate leaves £334" (NAAB Codrington Papers:A8).

Guinea corn (*Sorghum saccharatum*), also known as pearl millet, does well in arid environments like Barbuda (Kajuna 2001:6; NAAB Codrington

Papers:A5). It was introduced as a direct result of the Atlantic slave trade to Barbados in 1630 as part of a group of foodstuffs from Africa that included yams (Watts 1994:161–162). The monotony of Guinea corn was not lost on traveling physician George Pinckard (1816:301) during his visit to Barbados, when a slave remarked "me no like for have him guinea corn always! Massa gib me guinea corn too much. Guinea corn today! Guinea corn to-morrow! Guinea corn ebry day!" Such tedium could be offset with fresh produce from gardens and Sunday markets. The first evidence at Betty's Hope for provision grounds is depicted as the "Negroe Ground" on the 1710 map, which comprised open fields that were European in design and concept but worked by enslaved labor (Pulsipher 1994:205; see chapter 3).

Some of the produce included plantains and African-derived root crops like cassava, yams, and taro, which were ideally suited to Antigua and Barbuda's arid climate, requiring little attention and providing high yields. Root crops also had the advantage of being underground in storm- and hurricane-prone islands like Antigua. In fact, yams became so vital that provision gardens became known as "yam grounds" (Carney and Rosomoff 2009:112–113).

A pedestrian survey in 2015 revealed that the "Negroe Ground" was less than a mile south of Betty's Hope. By 1755, however, the provision ground was turned into pasture, and smaller house gardens replaced it, as indicated by Betty's Hope manager Joseph Walrond, who stated that slave gardens "should be allowed for ? of herbs or other little immediate supply of vegetables" (NAAB Codrington Papers:P10, C15; see Pulsipher 1994:203).

Trees were important in slave gardens because they provided shade and food. Typical trees included almond, papaya, breadfruit, and tamarind (Pulsipher 1994:214). This is evidenced by the current location of several mature tamarind trees in the slave quarters to the north of the great house at Betty's Hope. Other supplemental foodstuffs utilized local resources. Archaeological evidence in the form of a fishing weight, shells, and fish bones from the quarters' excavations reveals the utilization of nearby marine resources to supplement the slave diet (see chapter 6). Wallman (2014) reports access to marine sources at the Crève Coeur site on Martinique, where enslaved people utilized coastal and mangrove resources, similar to Antigua's environment.

Unfortunately, access to marine foods and small gardens may not have been enough to offset a deficient diet. Nutritional deficiencies in thiamine and vitamins A and B were common among the enslaved (Kiple and Kiple 1980:204–206; Kiple 1984:93–94). Indications of diet and disease among enslaved Africans require a multidisciplinary approach by combining historical accounts with bioarchaeology. In her analyses of 46 individuals recovered from Newton Plantation, Barbados, Kristina Shuler (2011:74) determined that the enslaved population was highly stressed. These individuals suffered from heavy workloads and died young. Although Shuler does not report on nutritional deficiencies from this sample, she did note a lack of specific disease markers, but she definitely observed infection in the lower part of the body (Shuler 2011).

The Codrington archives reveal that various maladies affected the enslaved population at Betty's Hope. Living conditions were a nightmare. Houses leaked in the rain and were routinely infested with rats, scorpions, spiders, centipedes, and lice (see Smith and Smith 1986:41–42). Such inhumane conditions were exacerbated by poor sanitation and exposure to microorganisms as people trod barefoot on plantation footpaths (Lawrence 2015:148).

Betty's Hope had its own doctor and sick house to attend to the enslaved population. Having medical care became mandatory in 1798 with the passing of the Amelioration Act, which was enacted to improve lives and maintain a healthy labor force (Gaspar 1993:113; Goveia 1965:195). According to Higman (1984:278), smallpox inoculations were first introduced to the Caribbean in the 1770s and were part of a broader plan of prevention to deter the spread of highly contagious and devastating diseases. Inoculation is mentioned in a letter to William Codrington II on May 19, 1780, from Betty's Hope estate manager Richard Clark, as he reports, "The Negroes upon the whole estates are at present healthy, there has been in the whole 61 innoculated for the small pox; in order to have the strictest care taken of them, I had them removed from the different estates to this, they are all in for a fair way except two, one of which has the appearance of some disorder on it and died thereof; another was suddenly seized with a violent vomiting and purging which carried it off" (NAAB Codrington Papers:C10).

Death's door was always close by, however. In another letter dated June 8, Clarke tells Codrington that "the number of Negros lost within the last two years has reduced the Gang" (NAAB Codrington Papers:C10). Some years before that, between 1777 and 1778, 80 slaves had died (NAAB

Codrington Papers:C12), although the cause of death is not mentioned. Some clues may be revealed in a letter dated December 10, 1779, from Richard Oliver to William Codrington, stating that "at Betty's Hope, 50 have died within 15 months . . . from a flue . . . of late" (NAAB Codrington Papers:C12). A litany of illnesses were recorded that included an outbreak of dysentery in 1778 and 1780, an earlier "malignant fever" in 1755, famine and disease in 1779, and a case of the dandy (dengue) fever in January through February 1828 (Antigua Almanac and Register 1843:82–87).

Perhaps one of the most prevalent diseases in the region was yellow fever. The wholesale destruction of forests for fuel and land clearing, exacerbated by stagnant water and a warm climate, created ideal conditions for mosquito-borne illnesses like yellow fever. Deforestation led to losses in bird species as natural predators of mosquitoes, while the importation of domestic livestock, soil erosion, and a host of other factors transformed island ecologies in less than 100 years (McNeill 2010:26–28). The main mosquito culprit was *Aedes aegypti,* the host of dengue and yellow fever viruses and more recently, chikungunya and Zika. Originating from Africa, this carrier of great suffering found a vulnerable population throughout the Caribbean. The claim that people of African descent had greater immunity to yellow fever has been challenged by Espinosa (2014).

In terms of infant and child mortality, Kiple (1984:120) maintains that factors contributing to low birth rates and infant and child death were a complex mix of genetic, nutritional, cultural, and immunological factors. There is little mention of infant and child mortality in the Codrington archive. Statistics for this and slave demography in general is a much needed area of study for Antigua and other islands for the colonial period.

Responses to diseases and ailments by the enslaved drew on local knowledge about herbal remedies. As Samuel Smith recalls (Smith and Smith 1986:64; see Handler 2000), "The bush syrup, for colds . . . was made up of mangy dagger, eucalyptus, sage and cattle tongue leaves along with bark mixed up with sugar. . . . To clean the blood you would take and boil white head broom and law lavington bush together with bitter mint and inflammation bush."

Our Ancestors Lived in Fear: Plantation Relations

In 2013, during interviews with local Antiguans, one couple remarked that "our ancestors lived in fear." Insight into the dynamics between planters, managers, and the enslaved population can perhaps shed light on this aspect of plantation life at Betty's Hope. From early on, one of the most pressing issues was the matter of the absentee landlord. Initially, in the first one or two generations, plantation owners like Christopher Codrington II and III actually lived on site, but subsequent generations bailed. The effects of absenteeship can be seen through Nevis planter William Stapleton, in what Keith Mason calls "involuntary transatlantic entrepreneurs." Their long-distance involvement positioned them in a vacuum, where "they combined vestigial patriarchal values with full-blown exploitation in an inherently unstable and potentially combustible cocktail" (Mason 2013:80).

What this meant for the enslaved at Betty's Hope and elsewhere was the inherent instability created through on-the-ground managers following directives given by their much-removed owners. When Stapleton eventually visited Nevis, he developed a deep suspicion of his new manager, "Herbert," over plantation expenditures, thereby creating anxiety and fear among his slaves. They reacted by absconding with mules at night (Mason 2013:86).

The Codrington correspondence reflects similar situations, as some owners got along well with some managers while others did not. One Betty's Hope manager in particular, Samuel Redhead, caused no end of headaches for William Codrington III during the 28 years of Redhead's oversight, from 1751 to 1779. Although highly adept, Redhead had his failings. In addition to his cruelty toward the aged and weak, Redhead had a rather liberal attitude toward enslaved women, siring a number of offspring in the process. He eventually settled with Sarah Bullock, with whom he had a son (Dyde 2000:88; NAAB Codrington Papers:CC36; see chapter 4).

The Redhead incident also stresses the vulnerability of enslaved women as special targets of unwanted male attention from planters and managers. A law passed in 1672 prohibiting such relations also stipulated that children produced from these unions would remain enslaved for life (which was not always enforced). Christopher Codrington II was particularly known for his libertine ways among enslaved women, from whom he produced a number of offspring. In his will dated July 15, 1698, Codrington compensated several enslaved women with freedom and money, among whom was Maudlin Marianus, with whom he had a son for whom he arranged an education.

Compounding matters, accusations of stealing were common, as was blame for attempts at poisoning, feigning illness, and plotting to overtake the island. For the sugar aristocracy, enslaved people were essentially not to be trusted. During William Codrington's oversight, his distrustful attitude can be surmised by a letter to his attorney, Slingsby Bethell, dated August 27, 1721, in which William laments his mother's stay at Betty's Hope:

> I hear my mother has got my footman Wm Lucky with her and that he lies in ye best chambers and eats and drinks with her. I beg you'l whip him out of ye house or else I shall expect to hear he has robb'd her and cut her throat for he is a vile villin as ever lived but if he was not I don't think she ought to entertain him or suffer him to be in ye house and I beg you'l not suffer him to be in ye house for if she should . . . I am sure he would steel and rob me of all my plate and jewells she has much. (NAAB Codrington Papers:C3)

One preoccupation was the issue of runaway slaves. Slave resistance on Antigua is recorded as early as 1666, when some brave individuals escaped to the mountains. The consequences of capture—surrender, recapture, floggings, or even death—were formidable. In the early years, there were still areas in which to hide, but deforestation diminished this possibility. In 1680, Antigua's Assembly passed the Restrayneing and Runnawayes Act, which rewarded quick recoveries. Those who escaped were not safe, since bounty hunters and militia were appointed by the island legislature to hunt them down. Legislation continued along these lines in 1697, 1702, and 1723.

In a letter to Edward Codrington from attorney Thomas Martin, dated January 25, 1773, Martin states that "I am informed that a Wench who was a Runaway a poor creature who was at Betty's Hope went to the mountains for change of Air and to [?] died last week" (NAAB Codrington Papers:C9). In another instance, William Codrington II was notified by his attorney on January 23, 1781, that "cash [was] paid out for bringing home a Runaway Negro named Isaac" for 8s, 3p. Five months later that same year, a runaway named "Greenwich" was returned for £8, 11s, 9p" (NAAB Codrington Papers:A5).

For those who chose not to run away, small daily acts of defiance were carried out, such as misplacing tools (Singleton 2001:108). One such instance that could be interpreted as resistance is recorded in a letter dated

April 14, 1827, from Betty's Hope Plantation manager John Osborn, where he states that a "Public enquiry [is being] held into the death of Codrington's cattle whose stomachs were found to contain pounded Glass, lead, copper, brass, pins, a piece of silver, [and] coarse sand" (NAAB Codrington Papers:CC344).

Excavations of the Slave Quarters at Betty's Hope

Taking all of this into account, excavation of the quarters at Betty's Hope was highly anticipated and critical for research. Locating the quarters was a long, slow process that required extensive surveys, whose data were integrated with satellite imagery overlays on the old maps. This vast area, covered in thick vegetation, held clues in the form of concentrated piles of rocks and tamarind trees (see Lenik 2005). Once the location was determined, excavations began at the edge of the area in 2014 and further continued into the heart of the quarters in 2015 (figure 10.1).

During the 2014 season, four 2 m^2 excavation units were laid side by side and excavated to bedrock (figure 10.2). The area turned out to be a large trash midden, whose assemblage included over a thousand pieces of Afro-Antiguan ware (see chapter 11), as well as annular and other imported Staffordshire wares, iron nails, discarded shell, and several items that relate to blacksmithing and equestrian activities (i.e., a stable; see chapters 2 and 6).

Moving deeper into the quarters in 2015, the excavation area comprised three 2 m^2 excavation units on a slope. The uppermost unit was fairly sterile, but mid-slope, the unit appears to have been a midden. The third lowermost unit had a rectangular stone foundation made of dressed stone, which was exposed at three corners (figure 10.3). The assemblage for the 2015 excavation comprised buttons, a thimble, buckles, bottle glass, clay tobacco-pipe stems, iron nails, and lead caulking, much of it fitting within the eighteenth century. Totals for both field seasons combined for imported ceramics (in sherds) include pearlware (681), annular ware (97), stoneware (94), whiteware (75), creamware (24), Delftware (8), yellow ware (6), and mocha ware (4) (figure 10.4). A similar assemblage for imported ceramics was reported for the eighteenth-century quarters at Schotsenhoek Plantation on St. Eustatius, thus challenging the notion that enslaved people had limited access to a variety of ceramic types (Stelten 2013:21–22).

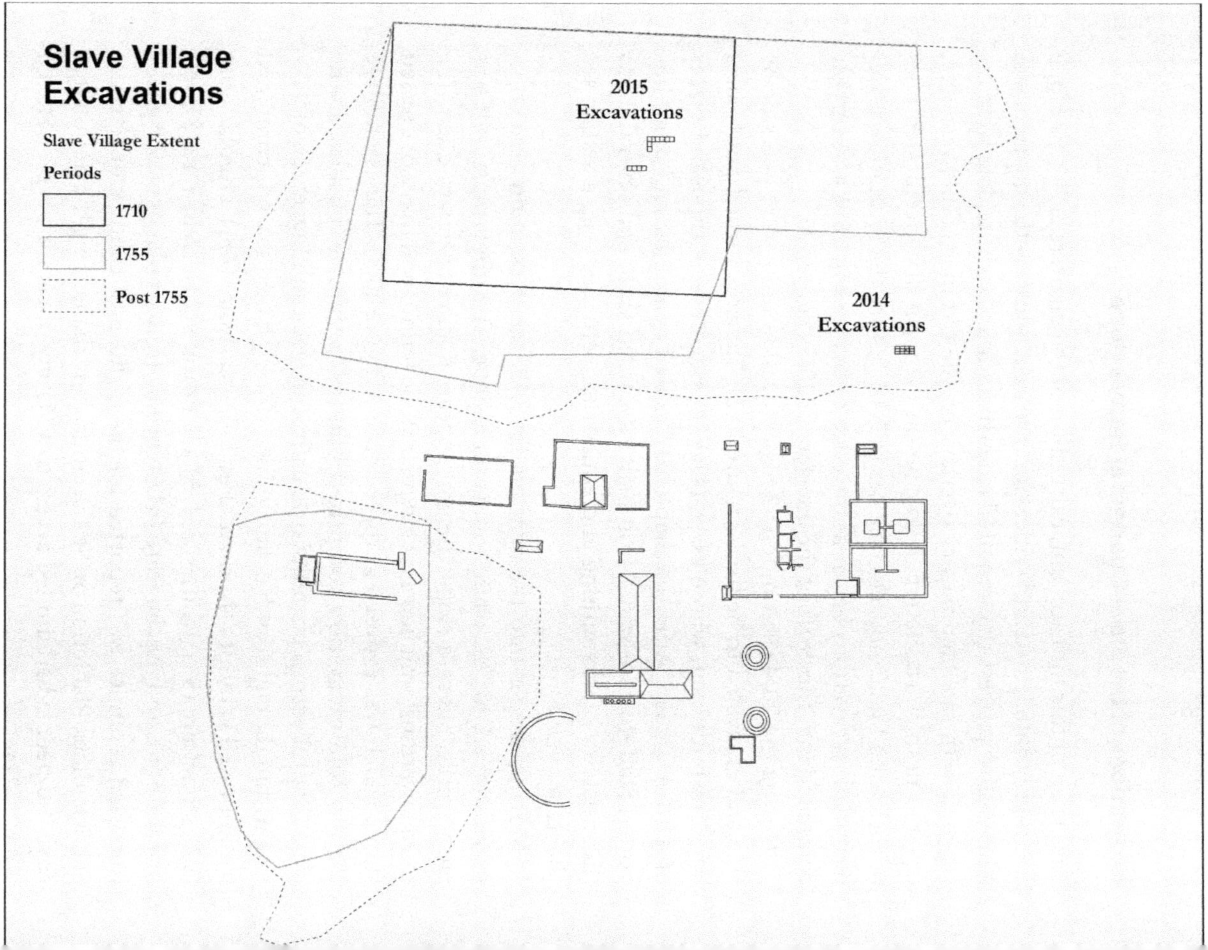

Figure 10.1. Areas of slave quarters excavated, 2014 and 2015. By

Figure 10.2. Area of 2014 excavations.

Totals for other artifacts from both field seasons include 82 buttons, 6 buckles, 15 beads, 21 pieces of flint, and 117 pieces of bottle glass, 21 of those medicinal. Datable clay tobacco-pipe stems were recorded in the following ranges: 1650–1680 (2), 1680–1720 (7), 1720–1750 (24), and 1750–1800 (71). By far the most ubiquitous artifact was the presence of 1,018 sherds of Afro-Antiguan ware, probably made by enslaved women at Betty's Hope (see chapter 11).

The discovery of the stone foundation poses some interesting questions regarding pre- and post-Emancipation housing and issues of identification.

Figure 10.3. Stone building feature from 2015.

Figure 10.4. Imported ceramics recovered in 2015.

Following Emancipation in 1834, some freed peoples decided to remain at Betty's Hope. Eventually, barracks housing was supplied for workers, creating some dilemma as to whether this stone foundation reflects the pre- or post-Emancipation era. The materiality and identification of quarters housing can be problematic, even for pre-Emancipation housing, as discussed by Espersen for the Dutch island of Saba (Espersen 2017:15). Yet in addition to the associated artifacts, there are a number of criteria to consider.

Earlier findings by Edith Gonzalez Scollard's excavations at Betty's Hope in 2002 can be instructive. The area where she excavated was about 40 m northeast of the Visitors Center, which lies at the periphery of the quarters area. During those excavations, two stone foundations were uncovered, one built of mixed stones of a vernacular style, the other constructed of dressed stone in a more formal style (Gonzalez Scollard 2008:81–82). The foundations of the more formal style suggest a type of housing that was meant to withstand tropical storms and flooding. That the building uncovered in 2015 is at the bottom of a slope helps substantiate this. Economically, this would also make more sense. Although slave housing is often characterized by free-standing wattle and daub structures, on Antigua, there appears to be a tradition of using stone foundations and walls, with thatched roofs made of cane trash (see Lanaghan 1991 [1844]:130; the stone foundation found in 2015 bears an ethnographic resemblance).

Most houses were small, no larger than 18 by 27 ft. (Higman 1984:220). The dimensions of the stone foundations from the 2015 field season measure 17.75 ft. wide by 33.66 ft. long, which is only slightly larger than published dimensions. Another argument for pre-Emancipation housing is based on barracks-type housing in the pre-Emancipation colony of Demerara-Essequibo in the Bahamas. Generally, such housing was abhorred by enslaved communities for its cattle-pen-like conditions (Higman 1984:221).

Finally, the reminiscences of Samuel Smith may shed light on the question by his description of housing at Betty's Hope being "about 60 feet long by forty feet wide and almost all the houses . . . [having] cellars made up of stones fastened together with white lime" (Smith and Smith 1986:41). These dimensions are more akin to post-Emancipation barracks-style housing rather than the smaller dwelling of 2015.

Housing quarters are rarely mentioned in the Codrington Papers. In a letter written between 1785 and 1792 to William Codrington II, Joseph Walrond described the "Negros Houses at Betty's Hope and the other plantations should by degrees are decaying, be built in regular sheets to prevent the general destruction by fire . . . [the houses] at Betty's Hope being huddled together" (NAAB Codrington Papers:C15). At some plantations, the houses were marked in rows or organized in some fashion; at Betty's Hope, it is difficult to determine original configurations based on one letter, also keeping in mind that such configurations could change over time. On the maps of 1710 and 1755, the arrangements are depicted as somewhat organized, but then the old maps are idealized.

Special Finds: Reflections of Agency and Resistance?

For Afro-Antiguans, there were few forms of cultural expression that could escape the watchful eyes of plantation managers. Three areas that can be considered possibilities for personal expression and some degree of cultural resistance include material culture, such as Afro-Antiguan ware (see chapter 11); religious practices, specifically in the form of obeah; and the Sunday market, which has been discussed extensively elsewhere (Bush 1990:46–50; Hauser 2008; Luffman 1789:65–66).

Cultural resistance, as shared practices among the enslaved, is reflected in their material culture. At Betty's Hope, special finds from the 2015 excavations include a lead fishing weight made from a repurposed musket ball, four cowrie shells (*Monetaria annulus*) from East Africa, cobalt-blue glass trade beads, a piece of wine bottle glass shaped like a projectile point, and a Delftware fragment shaped into a disk (figure 10.5).

The recovery of the four cowrie shells is especially powerful. Although archaeologists must be vigilant about being essentialist in interpreting certain artifacts as ethnic markers, such artifacts can provide a point of reference in analyzing identity (Singleton 2006:262). For example, Ryan Espersen (2017:147–148) reports a finding of a cowrie shell from Dutch Saba. Cowrie shells served multiple functions in West Africa, including "as currency, as decorations for clothing and hair, and as tools for divination" (Samford 1996:101). Cowries excavated in the colonial South were modified in ways consistent with many of these uses, suggesting that they served specific functions (Heath 2016). In 2015, work stopped so the Betty's Hope team could witness a magical moment when one of the cowrie

Figure 10.5. Special finds from 2015.

shells was recovered in the screen. The excavator who found the shell had lived in East Africa, so this was especially meaningful.

For the repurposed musket ball, bottle glass projectile point, cobalt trade beads, and Delftware disk, the aspect of reuse, although complex, is often a matter of making use of what is available (Wilkie 2016:337). By doing so, the beauty of it lies not only in its practicality, but also in its ingenuity. Bottle glass, as an artifact that is ubiquitous and plentiful at historic sites, offers multiple opportunities for re-creation, even when it can serve as decoration. Trade beads show up throughout the Americas in the living spaces of Africans, and they could have been used as status markers, for decoration, or as possessing special protective powers against misfortune and sickness, as documented in the American South (Samford 1996:102; Stine et al. 1996). The reuse of Delftware to make disks has been noted at similar excavations, including Potato Hill on Montserrat (Ryzewski and Cherry 2016:170–171). It has been suggested that these fascinating objects were gaming pieces or talismans or were used for trade (Armstrong 1990:137–138; Johnson 2012a, 2012b; Singleton 2001:110). Another disk made from hand-painted pearlware that appears as an eye was recovered as a surface find in the quarters, perhaps serving as a talisman.

Obeah

In addition to the material culture of resistance, the spiritual life of Afro-Antiguans was fairly much left to themselves. Planters were said to avoid pushing Christianity on their slaves because of their gaining access to literacy and not putting their "faith in massa." Although certain death rituals and customs are lost to obscurity, African traditions played a role in community life. After nine years of excavation at Betty's Hope, no African burial ground has ever been found. During the survey of the quarters area in 2013, however, a grave marker was discovered with an obeah sign on it.

The Obeah Act of 1904 made it illegal to practice obeah in Antigua and Barbuda, and it is still enforced (*Jamaica Observer* 2015). Derived from West African traditions, obeah invokes the world of spirits and supernatural forces, usually for personal use. As a form of resistance, obeah knowledge empowered community leaders and commanded respect as they possessed the wisdom to heal or harm, resolve conflicts, mete out justice, and practice a form of self-governance (Giraldo 2018; Handler 2000:59; Lightfoot 2015:146). To outsiders, obeah was witchcraft. This was especially the case after the post-Emancipation arrival of Christian missionaries on Antigua (Lightfoot 2015:186).

To anyone who has lived and worked in the Caribbean, the belief in ghosts and spirits of the deceased is well ingrained, such as the roving and unpredictable jumby on Antigua. At Betty's Hope, a carefully arranged set of bones surrounding a goat's skull with symbols drawn near them was discovered on the doorstep of an abandoned building. Betty's Hope is also no stranger to island beliefs about ghosts wandering the property. A little girl on a church picnic informed excavators that "there be jumby out here." Joy Lawrence (2015:151), an Antiguan writer and poet, reports local lore about workers at Betty's Hope who claimed to see ghosts posing as jack-o'-lanterns "who lured their victims by turning into a ball of fire and moving at great speed over hills and valleys." HAMAfilms Antigua, the award-winning film company owned by Howard and Mitzi Allen, has tapped into Caribbean folklore about the supernatural world, including in their 2011 production *The Skin*, partly filmed at Betty's Hope, in which a young couple discovers a cursed vase on the property.

11

Chemical Sourcing of Afro-Antiguan Ware from Betty's Hope Plantation

A Comparative Analysis

BENJAMIN C. KIRBY

Although much is known about the colonial British Caribbean, most written records pertain to upper-class Anglo colonists; information about the enslaved laborers who toiled on the plantation landscape is largely absent. Because the laborers left no records of their own, not much is known about them. This is especially true of the enslaved laborers who worked at Betty's Hope Plantation on Antigua. A primary reason for this lack of a written record comes from the strict slave codes of the colonial empire, which stripped slaves of their basic human rights and imposed punitive measures to repress and subjugate them.

Yet despite the imposed hardships, enslaved Africans managed to find agency and to engage in small daily acts of resistance. One way this was accomplished was through the production of their own pottery for everyday use in the form of gray-bodied Afro-Antiguan ware. This chapter therefore explores the Afro-Antiguan wares recovered from the Betty's Hope excavations, most of which were found at the slave village area. Additionally, a number of other pieces of red coarse earthenware were excavated; these are likely industrial and relate to the sugar industry. It has been assumed that the Afro-Antiguan ware found was made at the site or close by. To determine this, samples of Afro-Antiguan grayware and other coarse red earthenware were submitted to the Archaeometry Laboratory at the University of Missouri Research Reactor (MURR) for neutron activation analysis (NAA). This chapter will discuss the results of the tests and the implications as they relate to the lives of enslaved Africans at Betty's Hope Plantation.

Ceramic Production and the Sugar Industry

Plantations like Betty's Hope could not function without coarse earthenware ceramics. Ceramics were paramount to the production of sugar and for supporting the enslaved laborers of the plantation. The importance of ceramics can be seen in their ubiquity across sites in Antigua and the Caribbean region and testify to their widespread use during the historical period. Fortunately, a third of the geological makeup of Antigua is clay based, and excellent clay sources are nearby (Descantes et al. 2009:148; Handler 1964:150; Hauser and Handler 2009). Therefore, enslaved potters did not need to travel far to collect their clay for the manufacture of earthenware. However, despite the ample clay sources nearby, there is a chance for the presence of earthenware from outside Antigua at Betty's Hope Plantation.

The presence of a ceramic industry at Betty's Hope is documented in a 1710 survey of the Codrington properties by James Porter, where in the southern portion of the property, a building labeled "The Pott [*sic*] Work" exists (Goodwin 1994:103). Undoubtedly, this pot works served the plantation and manufactured the utilitarian wares needed to support a large population, as well as the molds needed for the manufacture of sugar. Codrington properties elsewhere in the Caribbean follow similar patterns, notably in Barbados, where Consett plantation boasted a pothouse (Bennet 1958:2; Kirby 2015:17).

Research Objective and Field Collection Methodology

The primary research objective of this study was to determine whether the enslaved potters gathered their clay from local sources at Betty's Hope in a similar vein to the observations of Handler (1964) or collected their clay from outside sources. For the purposes of this study, Afro-Antiguan ware is defined as gray-bodied utilitarian ware produced by enslaved individuals for their daily use, whereas redware likely comprises industrial sugar molds used in the production of sugar. Overall samples fall into two distinct categories: base clay samples in close proximity to Betty's Hope Plantation and ceramic sherds found archaeologically in the context of the slave village excavations at Betty's Hope. Of these two broad categories, the latter can be divided into three groups: Group A, which includes all typically identifiable Afro-Antiguan grayware from the slave quarters excavations that are characterized by a dark "burnt" interior paste; Group B, which are possibly non-typical Afro-Antiguan ware or imported coarse earthenware identifiable by a brown-colored paste; and Group C, jar sherd redware with large rims used in sugar production, either made locally or made elsewhere and imported.

The choice to focus on a collection of base clays from Betty's Hope stems from Handler's (1964) observation that enslaved artisans did not travel far to procure the raw clay necessary for the production of ceramics. Potworks Dam received a further emphasis, which served as a source of ceramics and raw clay during the historic period and existed in close proximity to the plantation (Museum of Antigua and Barbuda and Dockyard Museum 2012). The location of Potworks Dam is marked on the map of 1710 and is only a 15-minute walk from Betty's Hope (see chapter 10).

Admittedly, the geochemistry of raw clay can produce mixed results, namely that the chemical signature of a single clay deposit can vary tremendously over a small area. The primary reason for the inclusion of raw clay samples in analysis is to explore any possible relationships between base clay in close vicinity and those comprising finished ceramics. While unlikely, if a pattern exists between the raw clays and ceramics, it would be useful to any further study.

Samples of base clays from Betty's Hope were taken during the summer field season of 2015 after a survey of the area revealed suitable, or "good," sources of clay for extraction, focusing on the Top Road of the plantation located due south of the iconic double windmills (figure 11.1). "Good" denotes clay that is relatively homogenous, has little to no inclusions, has good elasticity, and is readily workable. Although Betty's Hope has an abundance of readily workable clay throughout the plantation, a recent drought on the island made the collection of clay difficult. Collection of samples was done by peeling back the first few centimeters of sunbaked clay resulting in a belt of moist clay a few centimeters below the surface. Fear of contamination from external sources prohibited the surface collection of clay. Altogether, nine sources were taken from Betty's Hope Plantation.

Collection at Potworks Dam followed a similar procedure, with a survey of the surrounding area and subsequent collection of samples. Unlike

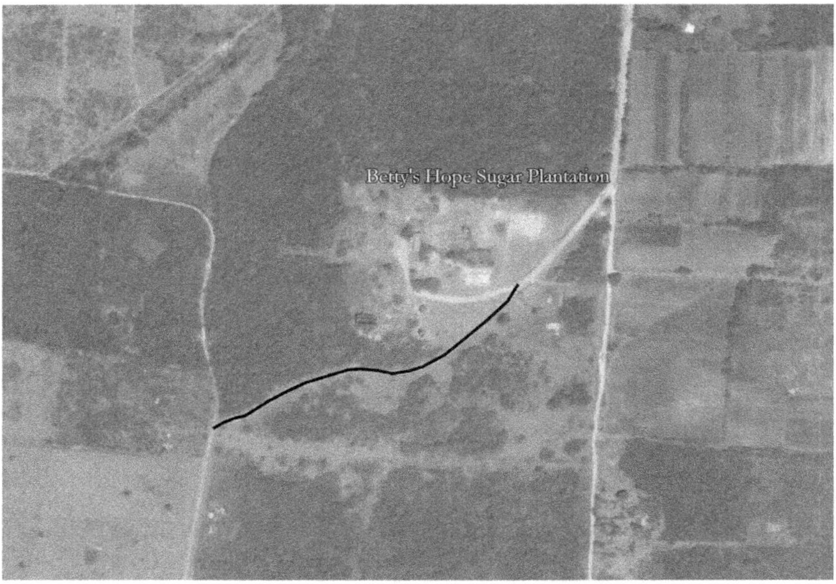

Figure 11.1. Location of the "Top Road."

at Betty's Hope, problems associated with drought were not encountered. Following the established protocol, the first few centimeters of moist clay were peeled back and the subsurface clay was collected. Samples were taken over a wider area than at Betty's Hope Plantation and mainly followed the perimeter of the reservoir. Potworks Dam yielded six samples, which are visually consistent with one another as well as with the samples taken from Betty's Hope Plantation.

Samples of pottery recovered from the slave village excavations were also collected. Three sample categories are present and descriptions are as follows: A (Afro-Antiguan ware sherds from a midden context, n = 41); B (brown coarse earthenware sherds, n = 31); and C (jar sherds with large rims, possibly involved in sugar production, n = 42). As such, a total number of 114 ceramic samples were included for analysis, as shown in table 11.1.

Other potential sources of ceramics for comparison exist that could expand the sample pool for this study, namely those of Armstrong, Galle, and Heath. According to Michael Glascock, MURR has access to these ceramics for potential inclusion in this study, but the samples either have not been submitted or analysis is currently under way (Michael Glascock, personal communication, February 12, 2016). However, during

Table 11.1. Sample Categories, Context, Description, and Sums

Sample Category	Context	Description	Sum (N)
CERAMICS			
A	Afro-Antiguan wares from a midden context	Grayware	41
B	Non-typical Afro-Antiguan ware	Grayware	31
C	Jar sherds with large rims, possibly involved in sugar industry	Redware	42
BASE CLAYS			
Betty's Hope	Clay samples from Top Road of Betty's Hope	Moist, gray, elastic, little to no inclusions	17
Potworks Dam	Clay samples from Potworks Dam	Moist, gray, elastic, little to no inclusions	13

interpretation of the data, these studies, if completed, can be included in the final sample pool for this study. As of this writing, MURR still does not have access to these data sets, and as such, they are not included in the analysis.

The sampling strategy was chosen to investigate any possible anthropogenic relationship between Betty's Hope and Potworks Dam and to address the research objective for this study. Raw clay samples were included in this study not as a method of identifying the exact location of raw material procurement, but instead to investigate any possible connection between the local sources and the composition of the sampled ceramics. A possible chemical signature match can prove useful in the interpretation of the data and could greatly aid in understanding where enslaved potters procured their clay.

Although plantations generally had ceramic production facilities on site (Goodwin 1994:103), oftentimes the lack of skilled laborers required procuring wares from outside sources, usually in close proximity and local, such as the Pothouse in Barbados (Bennet 1958:2). Potworks Dam, in close vicinity to Betty's Hope, was likely a secondary source of ceramic production for the plantation.

If the chemical signatures of the ceramics sampled correlate into two or more discrete groupings, then it is plausible that two different sources of clay exist. Although chemical signatures of base clay samples can vary drastically over a small area, strong and discrete groupings of ceramics can imply the presence of multiple sources. Comparisons drawn from the data in this manner could suggest an underground free exchange of goods with enslaved laborers on other plantations. Comparison of ceramics found at Betty's Hope to other plantations across Antigua could support this possibility.

Afro-Antiguan ware is distinct in its appearance in comparison to other coarse earthenware throughout the island. Afro-Antiguan ware can be consistently identified by a dark interior, and it may be glazed or unglazed. However, in the context of a midden that contains many broken sherds of coarse earthenware, identification of smaller pieces is not always conclusive, with locally made and imported sherds being mixed together. Ideally, the chemical signature of the ceramic outliers (not part of the majority grouping) could be instructive in this regard. Furthermore, analysis would also help us better understand the sourcing of the ceramics, as the discrete groupings would represent multiple sources for the ceramics.

Other studies have been performed analyzing Antiguan ceramics, namely Descantes et al. (2009). In their study, the ceramics of five prehistoric Antiguan sites were examined to better understand their relationships to one another. Conclusions from this study show that the bulk of ceramics analyzed belong to the same compositional group, implying the use of the same clay recipes and sources. While this study shows a strong correlation among the five groups of ceramics analyzed, it does not characterize ceramics from the historical period. The study of Descantes et al., however, demonstrates that geochemical analysis of Antiguan ceramics can produce valuable results.

Sample Processing, NAA, and Internal Groupings

Neutron activation analysis is an instrumental technique useful in determining concentrations of elements within a sample and results in both quantitative and qualitative data about a specimen. As with many instrumental techniques, NAA requires a standard of known chemical content to compare against the unknown sample. The National Institute of Standards and Technology (NIST) provides these standards that contain known concentrations of each element of interest. NAA centers on the creation of isotopes of each element and measures the radioactive decay of each resulting atom. Isotopes are atoms of the same element that have different numbers of neutrons in them, making their weight and chemical behavior different from the original atom.

The first step of NAA is to bombard the target nucleus with neutrons, which in turn creates an isotope of that element. The resulting isotope is usually radioactive, thus very unstable, and decays rapidly while emitting gamma rays (high-energy radioactive energy) and beta rays (high-energy radioactive electrons). A detector collects the emitted gamma rays and analyzes them based on their energy. In general, each gamma ray has an energy value that is specific to the element and when counted, quantifies the concentration present in the sample. Multiple energy levels of gamma rays can be produced from one irradiated sample, which yields a complex emission spectrum. Once all the gamma rays from a sample have been collected, the instrumentation analyzes the results and produces chemical data (Rouessac and Rouessac 2007:432–433). Samples were processed and analyzed at MURR, and Jeffrey R. Ferguson and Brandi MacDonald generated a report.

First, fragments of about 1 cm^2 were taken from each ceramic and abraded to remove any surface contamination, such as slip, paint, glaze, and adhering soil (Ferguson and MacDonald 2017:1). In this way, the compositional analysis of the specimen would reflect the chemical composition of the ceramic body itself. The exclusion of decorative elements and compositional qualities of surface enhancements (e.g., glaze and slip) is important to this study. Any applied glaze or slip could represent the result of local production; alternatively, it could also represent the import of material and application to the ceramic after bisque firing. The additional variable of importation of decorative surface elements is beyond the scope of this chapter.

Next, clay samples were fired in an oven and ground to a powder to homogenize the samples and ensure a uniform matrix of ceramic powder (Ferguson and MacDonald 2017:2). Two samples were prepared for each clay and ceramic specimen for short and long irradiation testing. Additionally, standards were created using NIST-certified reference materials (Ferguson and MacDonald 2017:2). The clay and ceramic specimens were then irradiated, and three counts of gamma rays were collected (table 11.2).

It is clear that many elements are considered when analyzing the chemical composition of ceramics, and analysis of all of them is sometimes fruitless. For this reason, various factors can exclude elements from

Table 11.2. Count Number, Irradiation and Decay Times, and Elements Examined

Count Number	Irradiation Time	Decay Time	Elements Examined
First ("Short")	720 seconds	1 day	Aluminum (Al), Barium (Ba), Calcium (Ca), Dysprosium (Dy), Potassium (K), Manganese (Mn), Sodium (Na), Titanium (Ti), Vanadium (V)
Second ("Middle")	1,800 seconds	7 days	Arsenic (As), Lanthanum (La), Lutetium (Lu), Neodymium (Nd), Samarium (Sm), Uranium (U), Ytterbium (Yb)
Third ("Long")	8,500 seconds	21–28 days	Cerium (Ce), Cobalt (Co), Chromium (Cr), Cesium (Cs), Europium (Eu), Iron (Fe), Hafnium (Hf), Nickel (Ni), Rubidium (Rb), Antimony (Sb), Scandium (Sc), Strontium (Sr), Tantalum (Ta), Terbium (Tb), Thorium (Th), Zinc (Zn), Zirconium (Zr)

analysis. For this study, the concentration value for nickel (Ni) was beyond detection limits and led to its exclusion from the final analysis (Ferguson and MacDonald 2017:3). Although many of these elements are trace elements, which I define as elements that are not frequently found in clay or geological formations in large abundance, they are useful in chemical compositional analysis. In effect, these elements are rare enough that their concentration throughout the earth's crust varies to such an extent that they are diagnostic of specific areas.

Using trace elements ties into the provenance postulate of Weigand et al. (1977:24), which states "that there exist differences in chemical composition between different natural sources that exceed, in some recognizable way, the differences observed within a given source." In other words, if a sample has a statistically significant commonality with a known source, it is highly likely that the sample originates from that source.

After identifying all the statistically significant elements, the raw counts were normalized using base-10 logarithms to create a normal distribution of the trace elements. This normalization fixes shortcomings of principal component analysis (PCA). According to Ferguson and MacDonald (2017:4), "'It is well known that PCA of chemical data is scale dependent' (Mardia et al. 1979), and analyses tend to be dominated by those elements or isotopes for which the concentrations are relatively large." By transforming the raw counts to a more normalized distribution (using base-10 logarithms), the tendency for elements such as iron (Fe) is largely eliminated, as iron is the fifth most abundant element in the earth's crust and found in large concentrations in soils (Haynes 2014:18).

In simplest terms, PCA is a type of multicomponent and factor-reduction analysis. A series of variables are taken into consideration and then reduced to the principal components that best characterize the internal variation within a data set. Plotting these principal components against each other causes internal groupings in the data between the components to manifest. Principal component analysis for this study resulted in the discovery of nine new ceramic compositional groups and three new clay groups, with respect to the MURR database (Ferguson and MacDonald 2017:2). Furthermore, some of the groupings showed favorable comparability with ceramics previously analyzed on Antigua (Descantes et al. 2009) and Jamaica (Hauser 2008; Kelly et al. 2008). The groupings that matched with the former case were a more consistent match and therefore integrated with the old data from Descantes et al. (2009). The groupings from the former case were not as consistent, and therefore the groupings were compared against each other rather than integrated (Ferguson and MacDonald 2017:6). What follows are brief descriptions of the groupings as discovered by MURR; a more in-depth discussion of the groupings will occur in a later section of this chapter.

MURR Groupings

Group 1 consists of 10 samples from Betty's Hope (around 9% of the total specimens) and has very different chemical compositions from the rest of the samples analyzed (see table 11.3). This grouping matches up nicely with two previously discovered groupings from Saint Anne's and Saint Andrew's parishes on Jamaica. Furthermore, this grouping contains the majority (~63%) of the redware analyzed for this study. While no definitive match can be made between the ceramics in Group 1 and those of Jamaica, the close chemical similarity suggests that the ceramics were produced at an unknown manufacturing site and then imported to both

Table 11.3. Groupings, Chemical Characteristics, Sum of Samples, and Source Location

Grouping	Chemical Characteristics	Sum of Samples	Source
Group 1	Elevated concentrations of almost all minor and trace elements	10 of 16 redware sherds	Not from Antigua, similar to ceramics from Jamaica
Groups 3/4/5/7 and CANT3	Higher iron concentrations and manganese concentrations	Only samples from Descantes et al. (2009) study	From Antigua, local chemical signature
Group 6	Lower iron concentrations	11 grayware (Afro-Antiguan) sherds	From Antigua, similar to Group 2, variation of main local ceramic production signature
Group 9	Different concentrations of rare earth metals	2 grayware (Afro-Antiguan) sherds	From Antigua, most are from Descantes et al. study
Groups 2/8	Higher iron concentrations	47 grayware (Afro-Antiguan) sherds, 6 redware sherds	From Antigua, local chemical signature
Unassigned	Don't fall statistically into other groupings	3 grayware (Afro-Antiguan) sherds	Unknown, signifies standardization in ceramic industry

islands. This is not to say that there is a direct exchange between the islands, but merely that the unknown production and procurement site is related in some way to both islands (Ferguson and MacDonald 2017:6).

Groups 3, 4, 5, 6, and CANT3 (a grouping discovered from the Descantes et al. [2009] study) show elevated iron concentrations (which were further separated by their respective manganese [Mn] concentrations) and only contain the prehistoric samples from Descantes et al. (2009). This grouping seems to represent one of the chemical signatures of Antigua, and because it contains only Amerindian ceramics, a safe argument is that most of these represent locally produced ceramics. None of the ceramics from this study of Betty's Hope appear in this grouping, however (Ferguson and MacDonald 2017:6). Additionally, the grouping from Descantes et al. (2009) that does not match up with the rest of Antigua's chemical signature does not appear in this grouping.

Group 6 is characterized by lower iron concentrations and consists of 11 (~11% of the total) grayware ceramics from Betty's Hope. Overall, this grouping is similar in chemical composition to Group 2 (the main grouping for Antigua) and most likely represents a slight variation in the local ceramic production signature (Ferguson and MacDonald 2017:6).

Group 9 consists of two ceramics from Betty's Hope, while the other 18 samples are from Descantes et al.'s (2009) study of terminal- and post-Saladoid ceramics. While there is no visual distinction between the ceramics of Group 9 and Group 2, statistical distance formulas (i.e., Mahalanobis distance projections) help to separate the two groups (Ferguson and MacDonald 2017:7).

Groups 2 and 8 contain 62 specimens, 53 from this current study and nine from Descantes et al.'s (2009) study on pre-Contact ceramics. Groups 3, 4, 5, 6, and CANT3 are subdivisions of this grouping, and Group 6 is similar enough to be a variation of this grouping. Group 2 is a tighter cluster within Group 8, and together they represent the main chemical signature of Antigua. Combining these two groupings together to make a main chemical signature is a safe bet and shows the chemical signature for locally produced ceramics. Furthermore, the inclusion of the Descantes et al. (2009) specimens adds to arguments that this is the main chemical signature of Antigua (Ferguson and MacDonald 2017:7). Interestingly, of the 53 specimens from this study included in this group, six are pieces of redware that are likely connected to the sugar industry.

Finally, the rest of the ceramics studied (~3%) fall into the unassigned grouping. The low number of unassigned ceramics breaks the norm in chemical composition studies. According to Ferguson, generally, 20–30% of unassigned ceramics are expected in studies; however, the low number of unassigned ceramics is likely "due to a combination of mainly industrial ceramic production," where there is a more standardized preparation of the clay. The three ceramics that were included in this grouping from Betty's Hope were all Afro-Antiguan grayware (Ferguson and MacDonald 2017:7).

Also included in the study were 30 clay samples from two likely historical sources: Betty's Hope Plantation and Potworks Dam. Unfortunately, analysis of raw clays rarely yields concrete relationships to archaeologically recovered ceramics; however, there is clear patterning evident within these clay samples. Three groupings were identified, as follows: Clay Group 1 includes nine samples all from Betty's Hope Plantation; Clay Group 2 includes eight samples from Betty's Hope Plantation and one from Potworks Dam; and Clay Group 3 includes 12 samples all from Potworks Dam. While the clay groupings do not match up well with the

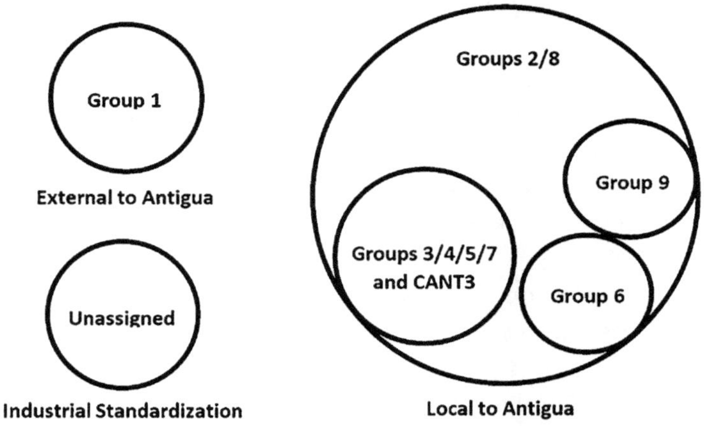

Figure 11.2. Visual representation of groupings.

ceramic groupings (with each ceramic from Groups 2 and 8 having a 0.05% chance of membership), the data can be forced to fit to compare to the clay groupings.

When the data of the clays and ceramics are both transformed, some similarity between Clay Groups 1 and 2 and Ceramic Groups 2, 8, and 9 arises. Analysis of the data suggests some patterning between the base clays from Betty's Hope and Potworks Dam, but it is not statistically significant enough to definitively match the ceramic data to the base clay data (Ferguson and MacDonald 2017:7–8). This is not entirely surprising, since as previously stated, the chemical content of base clays and ceramics rarely match up. A more comprehensive sampling of the island would explore this idea further.

The results from MURR show several interesting aspects of the data set with respect to the Afro-Antiguan grayware (utilitarian) and redware (likely sugar molds) found at Betty's Hope. The chemical data suggest three main clusters of the groupings identified from MURR: external to Antigua, local to Antigua, and evidence of industrial standardization. Of these three main clusters, the majority of the ceramics fall into the locally produced category (see figure 11.2).

The majority of the Afro-Antiguan grayware at Betty's Hope has chemical signatures that suggest local production on the island. Groups 2 and 8 represent the primary chemical signature for Antigua, and ceramics that fall within this grouping definitively represent local production. Within this main chemical signature, several variations of the signature were found: Group 6; Group 9; Groups 3, 4, 5, and 7; and CANT3. Groups 6 and 9 are composed entirely of Afro-Antiguan grayware and fit the hypothesis of local production of utilitarian wares at Betty's Hope. Groups 3, 4, 5, 7, and CANT3 further reinforce the notion of local production, as all the ceramics in this grouping come from the Descantes et al. (2009) study on pre-Contact ceramics. It is important to note that this grouping does not include the externally sourced ceramics from the Descantes et al. (2009) study, but just includes the locally produced ones.

The analysis of the redware proved to be very interesting, as there is a split between local production and external production. The majority of the redware pieces (~63%) fall into Group 1, which has a different chemical signature from the Afro-Antiguan grayware and matches up nicely to ceramics from Jamaica, which suggests a common source between the two islands. This would largely make sense, as most sugar molds would have been imported from abroad. However, six redware sherds have chemical signatures that include them in Groups 2 and 8, the main primary chemical signature of Antigua. This suggests that they were both locally produced and produced at the same site as the other 47 Afro-Antiguan pieces of grayware in the grouping.

The last cluster is the three unassigned sherds that did not fall neatly into the other clusters. While it is typical for 20–30% of total specimens to fall into this category for similar studies, the relatively low percentage of the total sample population suggests heavy standardization of the ceramic industry for Betty's Hope.

Discussion: The Historical Record and NAA Results

From the historical record, it is clear that the enslaved laborers sourced clay locally (Handler 1964:150; Descantes et al. 2009:148) and that ceramic production was present at Betty's Hope (Goodwin 1994:103). These enslaved potters would have produced the utilitarian Afro-Antiguan gray-bodied coarse earthenware pieces central to their lives. The chemical data from the NAA analysis done by MURR reinforces this notion, as all grayware from the study shares a common source. This means that the clay was sourced from the same area and the ceramics themselves were produced in the same place. This is consistent with the presence of a pot kiln at Betty's Hope and is not unheard of for plantations in the Caribbean.

The ability of the enslaved potters to choose their clay sources demonstrates a high degree of agency. The enslaved potters would have made a deliberate choice of where to source their clay and would have been allowed to leave their normal boundaries on the plantation to procure it. The exact location of their source is hard to pin down through the chemical data. While clay samples were taken at likely sources of clay (Betty's Hope Plantation and Potworks Dam), it is difficult to match up the chemical signatures of the base clays with the Afro-Antiguan grayware and industrial redware. Furthermore, the entire ceramic industry of Betty's Hope seems to be highly standardized, as demonstrated by the low degree of variability in the chemical characterization data. As such, the enslaved potters had a high level of control over all aspects of the production process.

Concerning the redware, which represents industrial ceramics for the production of sugar, there seems to be a disparity between the historical record and the chemical data. Historically, importation of sugar molds was a cost-cutting measure. The chemical data suggests a common source between the majority of the redware at Betty's Hope and plantations in Jamaica. However, a good portion of the redware (~37%) is included in the primary chemical signature for Antigua (Groups 2 and 8), which begs the question, Why did enslaved potters produce redware at Betty's Hope?

There are two possible answers: the plantation had a need for sugar molds to be produced on site and/or the enslaved laborers were producing utilitarian ware for sugar production separate from their own Afro-Antiguan grayware. As Betty's Hope was the flagship plantation for the wealthy Codrington family, it was noted for a self-sufficiency that made it "Antigua's most efficient, large-scale sugar plantation" (Goodwin 1994:100–101). It appears, then, that the enslaved potters were branching out in their ceramic production and producing different ware types, including their own Afro-Antiguan grayware. This suggests the possibility that there is more variation in Afro-Antiguan grayware than previously thought. Samples of Afro-Antiguan cooking and water pots displayed at the Museum of Antigua and Barbuda indicate a variety in coarse earthenware made by Afro-Antiguans differentiated not only by function and shape, but also by differences in colors of paste. Therefore, coarse earthen grayware previously thought to have been imported and separate from more typical Afro-Antiguan pottery may in fact be non-typical variations in locally made Afro-Antiguan ware.

This also suggests that Afro-Antiguans were deliberate in where they sourced their base clay and in the standardization of the ceramic production for the plantation. There is already a high degree of agency implicit in these processes, and the production of a new type of pottery could be an expression of their innovations, working with local clay sources to possibly produce pottery for sugar production, everyday general plantation use, and their own distinctive Afro-Antiguan ware. A further study of the redware with a focus on function, form, and any decoration could explain this phenomenon.

While knowing the exact source of the base clay would be valuable to future studies, the chemical data of the base clay does not fit well with the ceramics. Although the base clays did group internally, it is simply not possible to draw concrete conclusions from this data.

Conclusion and Future Studies

The combination of chemical analysis and examination of the historical record can produce valuable insights into the past for groups that left no written record of their own. Chemical analysis of the ceramics from Betty's Hope shows that the enslaved potters of the plantation had a high degree of agency with regard to the ceramic industry. Additionally, the potters had complete control over all aspects of the production, from sourcing the clay to utilizing the ceramics. While the pieces of redware examined came mostly from an external source and originally were thought to be industrial and related to sugar production, some of them were produced locally at Betty's Hope. The locally produced redware could be industrial, but it could also represent an effort to create a unique community identity.

While this study yielded valuable results, future work could be constructive. Further sampling of ceramics across Antigua from a similar time period could yield valuable information on possible internal exchange networks between plantations. More information on this subject could further explain resistance among the various communities of enslaved individuals on Antigua. Similarly, a more exhaustive sampling of clay sources across the island could give a more definitive match between the base clays and the ceramics analyzed. Such a study would help track any internal exchange networks on the island. Currently, the base clay data does not match up well to the ceramic data, and conclusions between the two data sets are hard to draw concretely. Finally, a deeper exploration of plantation ware could help explain the local production of some redware at Betty's Hope.

Acknowledgments

I would like to thank the University of Missouri Research Reactor for analysis of the ceramic samples, and Jeffrey Ferguson and Brandi MacDonald for generating the final report that proved instrumental to writing this chapter. Furthermore, partial funding support was provided by NSF grant #1621158.

PART V

Betty's Hope and Beyond

12

"Primed with Flip and Toddy"

The Globalized Material Culture of Rum Production at Betty's Hope Plantation, Antigua

CHARLOTTE GOUDGE

The study of plantations and their sociocultural character often overshadows the study of intensified production in the Caribbean. However, given that the primary objective of plantations was always commodity production, this aspect of plantation operations is vital for achieving a holistic vision of plantation life. For the historical archaeologist, material culture studies are a keystone of academic inquiry that permits an all-inclusive and detailed understanding of human-object relations.

In the case of a commodity-based approach to the cultural history of rum, this includes an understanding of production, commoditization, and the use of commodity chains. Commodities like rum, and their production, distribution, and consumption, have become an integral part of modern socioeconomic studies. The cornucopia of commodities available to British society, their integration into the fabric of the early modern period life, and the intensification of consumption can be directly equated to modern consumption styles.

This chapter will focus on rum production at the Betty's Hope Plantation and how it related to large-scale productive and consumptive themes ongoing during the expansion of British markets in the early modern period. This will be done utilizing evidence gained through archaeological excavations conducted at Betty's Hope from 2013 to 2015 and a survey of the Codrington Papers archive. The evidence shows that rum was being produced from as early as 1710 (the earliest mapping of the plantation) in small separate buildings dedicated to each productive stage (boiling, curing, distillation). During the late eighteenth century, these buildings were eventually replaced by a much larger, integrated, T-shaped factory that included all three production processes.

To this end, my research employs a variety of multidisciplinary approaches to fully reveal the global, cultural, and economic narrative themes of production, distribution, and consumption that arise as part of this multifaceted study. This includes a study of the social and economic history of the period to understand the main themes and the geographical, social, and cultural paradigms of the period, including documentary research into methods of production and consumption of rum in the Caribbean and the wider world. This approach is used as an integrated research method into the social and technological changes that took place during the development of alcohol consumption in the early modern world and rum distillation in the Caribbean. This is particularly useful in situating rum within the development of the consumer revolution and the development of mass production. I have employed conceptual analysis of the topic, discussing the major theoretical approaches, to debate and evaluate the impact that rum had on production and the consumer. As Marx noted, "It is not the consciousness of men that determines their existence, but, on the contrary, their social existence that determines their consciousness" (Marx 1979[1859]:11–12).

Background

The impact of rum on Atlantic society has garnered little academic attention, with some exceptions (see Smith 2005). Yet along with other key commodities such as sugar, tobacco, and tea, rum's impact as a commodity was immense, specifically during the early modern period and the formation of globalized market systems. The commodity chain experienced by rum demonstrates a constantly fluctuating symbiotic relationship between changing social and technological themes felt during the production, distribution, and consumption processes.

Betty's Hope was the pioneering British plantation in Antigua, an island that remained resolutely British for all but six months of its European history and that is a primary example of a dynamic, changing social and productive Caribbean environment (for more on the history of Betty's Hope, refer to chapter 2). Evidence suggests that Betty's Hope was producing rum by 1710, the period in which rum production became if not commonplace, certainly increasingly popular. British desire for rum proliferated dramatically into the early to middle years of the eighteenth century.

In the British West Indies, rum imports spiked from 58,000 gallons in 1721 to 320,000 in 1730 (Williams 1994:78). The absorption of alcohol into British life only grew from that point, and a rivalry began between West Indian rum and the already well-entrenched Dutch gin, which was by now a very popular tipple for the British working classes. Imported to Britain in large quantities as a result of an abundant grain harvest in the Netherlands, gin became widely accessible and very cheap to consume.

Several groups interacted with rum. Illicit traders in West Africa used it as currency; gifts of alcohol, broadly known as *dashee*, *dash*, *dassy*, and *bizy* (Barbot 1732:142, 260; Smith 2005:96), were required by local dealers and chieftains prior to trading. West Africans alternately used rum as a luxury or religious product. Many Africans seemed to value imported alcoholic beverages for their newness, especially distilled spirits, which were much more concentrated, or "hot," than their usual fermented drinks. Rum soon became an elite commodity, valued by African leaders and royalty (Smith 2005:130) and absorbed within West African feasting and religious practices. Enslaved Africans similarly used it within folk religions such as obeah and vodou and as a platform for opposition tactics.

Theoretical Approaches to Material Culture and Rum Production

There are many types of production theory, a term that covers the economic practice of creating goods or services using available raw materials and resources. Different forms and styles of production can be depicted as reactions to the varying cultural needs developed by a society. These needs, whether for vital necessities or desire for luxury products, reflect the relationship between production and consumer demand. The process of creating a commodity—the transformation by cultural process of a raw material into an infinitely tradable good—is affected at every step of its operational sequence by various cultural factors affected by its housing society. *Chaîne opératoire* is a methodological tool used for analyzing the technical and social acts that revolve around the production, use, and eventual disposal of artifacts. In practice, the *chaîne opératoire* allows the archaeologist to examine the steps taken by the producer to produce the object. With this in mind, it is not unreasonable to utilize this method to study modern and early modern—as well as ancient—industrially produced materialities.

Even within the creation of an early modern product such as rum in the Caribbean sugar islands, the concept of style remains in flux. Production style in this case is dependent not only on the nationality of the actors engaged in production, but also on the materials and environment available to the specific Caribbean island and the producers themselves and the location and layout of plantation-specific production.

Recipes and production techniques varied from island to island. This isochrestic style represents the choices made by the individual producers and how their decision making affected idiosyncratic manufacturing styles and therefore regional micro styles (Sackett 1982). Technical variations in rum production methods can also be seen among Portuguese, Spanish, French, and English producers (Ligon 1673; Martin 1750; Wray 1848).

These sorts of cultural factors had large-scale implications on rum production, the repercussions of which materialized to force many producers, such as those at Betty's Hope, to implement new technological advances into the *chaîne opératoire* of rum production to survive. In turn, rum also had a clear influence on individuals. This can be seen in the intimate producers (the workers who physically create the material) and the stimulating producers (the landowners who stimulate and facilitate the production without directly creating the material). The material is also directly affected by the machinery utilized as part of a productive event.

By approaching rum as a socially influenced material, it becomes possible to validate and analyze a number of material cultural processes and effects. Commodity production relies on the "action, innovation and development of humans" (Goudge 2017:2). The consequences of the productive decision-making processes are therefore active both socially and physically. The corollaries of production are therefore often directly reflected on the materiality of production and its final product, in this case rum. In the pursuit of intensive and efficient production of rum, local and global societies were transformed.

Finally, the specific nature of the material culture in this case directly affected the personhood of not only those who created the material, but also those who consumed it, leaving an indelible social mark.

Rum and Betty's Hope

In the early years of the seventeenth century, rum was a waste product, subordinate to "king sugar." During this time, though a small amount was earmarked for export, most rum was produced for consumption on plantations. Specific features of the plantation physiognomy, particularly the availability of virtually limitless workforces of enslaved peoples, expedited the exponential growth of Caribbean production. However, the drastic development of consumer demand for luxury commodities (such as rum and sugar) in the eighteenth century also caused the need for the improvement and expansion of production methodologies (Goudge 2017). The development of industrial machinery beginning in the 1760s alleviated the pressures that had recently been placed on many plantations by increased consumer demand. Industrial advances that became commonly available to plantations during the eighteenth and nineteenth centuries were both a reaction to and a catalyst of social and physical change. The intensification of demand for commodities and the eventual abolition of slavery necessitated the modification of productive methodology and technology.

Consumer desire and impending technological advances forced a change in the layout and material culture of Caribbean sugar plantations. The corporeal construction of sugar plantations was a continually developing system. The topographical, structural, and financial influences of commodities existed within oscillating and lively structural and social landscapes of the Atlantic World. In the case of Betty's Hope, the plantation underwent continuous amendments to the production methodology and organization and the size of the buildings.

The original sugar works (boiling and curing houses) are now long gone, but maps and plans in the Codrington archives have been informative in determining the placement of the buildings in the factory complex. These have been instructional regarding the use of the factory (and the still house), the movement of product, and the changes that correlate with the plantation's growth and change.

The earliest map of the plantation, made in 1710 (NAAB Codrington Papers:P3), shows the first known depiction of the plantation's layout. This arrangement shows the still, boiling ("boyling"), and curing houses as three separate two-story buildings. The still and boiling houses were located to the southwest of the plantation, while the curing house, a much larger building (if the scale is to be believed), was north of the road out of the plantation situated with its own "curing house cistern" that held 353 tons. The still and boiling houses were both located close to the windmills, likely to optimize production and reduce traveling time and distance for the cane juice to be transported from one to the other.

It is probable that the topography of the plantation was utilized to simplify the transport of materials. The windmills are situated on top of the hill to capture maximum wind coming from the sea, while the production spaces (in particular the boiling house) were built lower down on the hill, which made the transport of the raw cane juice an easier endeavor. By the time the product was transported from the boiling to curing house, it would have been a less viscous material making it much more easily transportable.

Betty's Hope, it seems, retained this configuration for a number of years, as shown by successive maps produced in 1755 and 1780. Several versions of a map produced in 1755 are available in the National Archives of Antigua and Barbuda. All of these rendered drawings on rag paper illustrate various iterations of Betty's Hope and other adjacent Codrington properties. One copy appears to match an original, and a third seems to have been rushed, showing little detail and a different configuration of buildings. The presumed original 1755 map shows the plantation in much more detail than the others, again showing the three separate buildings in a similar configuration as in 1710. The major change shown is the altered location of the main road out of the plantation. In 1710, the main road in and out of the plantation was from the west. However, in 1755, the road comes from the east (as it does in the modern period) curving around to the boiling and still houses. This is an apt depiction of changing exports from the plantation, corresponding to changing locales for roads and wharves to transport these export goods.

The 1780 map, titled the "Ground plan of the Old Works on Betty's Hope in the Island of Antigua" (NAAB Codrington Papers:P9), depicts these early production buildings in greater detail. Interestingly, unlike the 1710 map, which shows both buildings as having two stories, this map shows each building as having a single story only. This map shows, in detail, the layout of the still, boiling, and curing houses.

The boiling house structure is shown to have had a large boiling train of copper boiling pans (or taiches), where the sugar syrup would have been reduced. The building to the east of the construction is described as the curing house, which is made up of not only the curing area but also a storage area and two unlabeled spaces.

The still house is made up of a number of structures. The stills themselves (and presumably the furnaces used to heat them) are situated outside, not uncommon for the period; the Wetherills plantation had a similar arrangement. The structure marked on the map as "D" is unfortunately unlabeled within the key; however, it seems possible that it was the fermentation vat. Cisterns are marked on the plan (F), and interestingly, it is noted that a worm coil, a form of condenser used within the distillation process to better separate the alcohol from the water, was in use.

The still house building has been divided in two, with the room to the left labeled as the rum house and the room to the right generically called the still house. The uses of these rooms are unclear; however, the rum house may have been where the casks of finished product were stored, whereas the still house was where distillation occurred.

The fermented liquid would have been transferred to the still to be heated and passed through the worm coil before being sent back into the building to be barreled. Pot stills were the most common form of still in this period. In 1715, it is noted in the Codrington Papers that there were four pot stills being used in the plantation (NAAB Codrington Papers:C2). A manufactory with this layout could produce at least 20,000 gallons of rum in one crop (NAAB Codrington Papers:C8).

It seems that at the time of this 1780 map, the factory was undergoing a radical updating of both structure and methodology. Comparison of the 1710 map to the later 1755 map (NAAB Codrington Papers:P10) and a notable mention in the Codrington correspondence suggests that an older still house was torn down and fully replaced by the new, improved factory works as plantation operations expanded by 1780.

The next depiction of the plantation's sugar factory operations, discovered in the Codrington archives, shows a drastically and fundamentally altered production landscape. The plan, signed by W. J. Blackburn and dated to 1862, illustrates a detailed rendering of the factory works (figure 12.1). The map illustrates a large, T-shaped, integral factory works structure containing all three productive spaces: the boiling, still, and curing houses. This configuration would have been designed originally to be

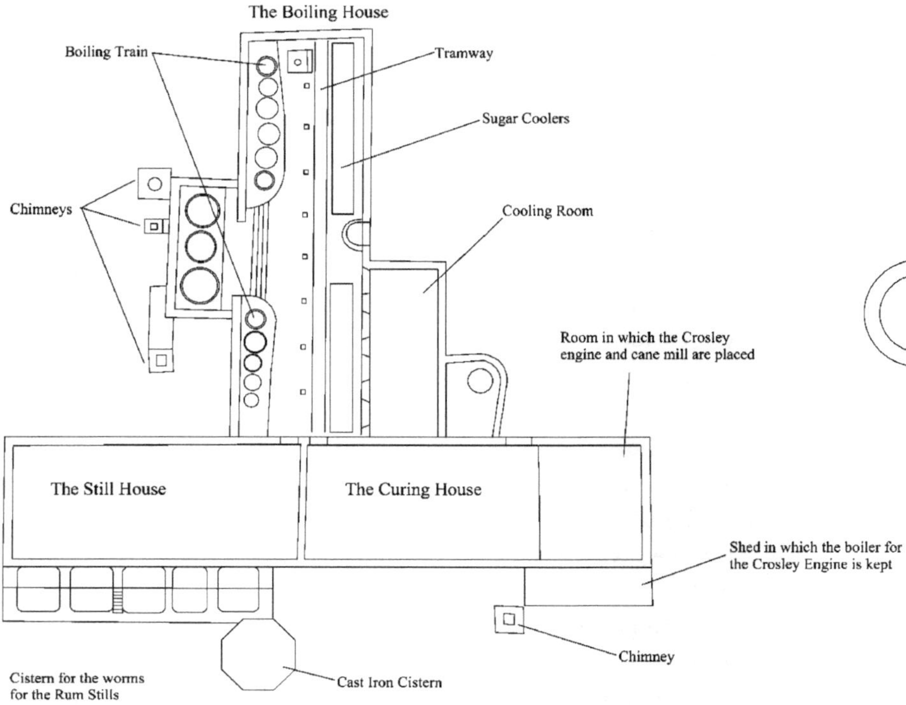

Figure 12.1. Plan based on the schematic of the Betty's Hope factory works, 1862. Illustration by Iona Duckworth and Charlotte Goudge.

used without steam, which was not experimented with on the plantation until c. 1845. Production methodologies would have been updated between 1845 and 1862 to include steam.

There is scant evidence to unpack the pre-steam methods used at the Betty's Hope still house. The 1862 schematic notes a "Still Room with Rum Cellar underneath"; therefore, unlike the boiling house, the still house section of the building was, in fact, a two-story structure with a rum cellar built into the side of the hill. This suggests that like at the Mount Gay Distillery, Barbados, the fermented liquor was piped up to the second floor and distilled in a pot still before being sent (utilizing gravity) to the worm coil in the cisterns (figure 12.2). The liquor would then be sent through lead pipes that go straight through the wall of the cistern and back into the cellar of the still house to be barreled or canned.

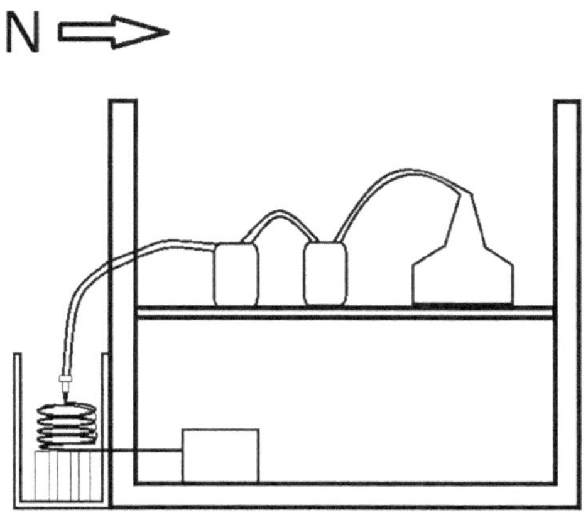

Figure 12.2. Diagram of hypothesized rum distillation at Betty's Hope.

The term "canning" is used to describe not only the use of metal cans, an invention that was patented in 1810, but also the use of what would be known today as crocks or jars. The large quantities of stoneware discovered at the BH-B site could be remnants of the stoneware crocks and jars used to transport rum, particularly for internal, island use.

Rum cellars were utilized for canning, aging, and storage, though Wray (1848) states that a great deal of rum was often shipped immediately rather than held in storage. This could be because the exportation process was long in duration, varying taste for rum "sharpness," or because of taxation and levy fluctuations. The Codrington accounts suggest (without being explicit) that when the purchase price was low, the rum at Betty's Hope would occasionally be held back at source until the prices became more financially favorable for the plantation (figure 12.3).

The advent of the industrial period, which began in earnest in the Caribbean between 1830 and 1840, instigated the adoption of steam-powered engines. The grinding of the cane within the now famous Betty's Hope windmills ceased at the introduction of steam (Reginald Murphy, personal communication, June 20, 2015), replaced by a steam mill that was kept in a specifically designated room on the eastern end of the curing house (see figure 12.1). The 1862 schematic repeatedly notes the name Crossley/Crosley in connection with this engine. In the Codrington Papers of 1844, there is also mention of a "Cambrian Engine" already being

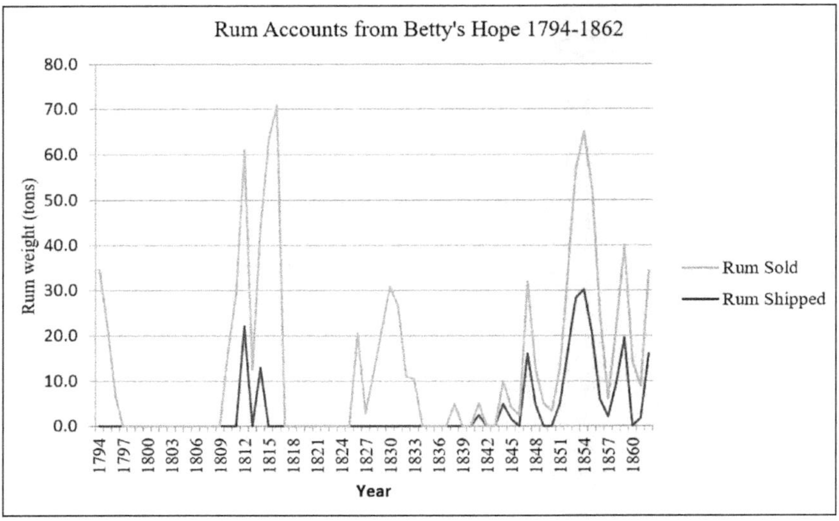

Figure 12.3. Rum shipped or sold from Betty's Hope, Codrington Papers Accounts, 1794–1862.

in place on the plantation (NAAB Codrington Papers:A64). The Cambrian engine was an early form of a non-condensing steam engine developed by one Henry Crosley, a civil engineer. There is very little literature on the configuration of steam-powered Caribbean still houses; yet still-house technology continued to advance with the development of new forms of distillation. This is particularly evident in the steam-heated pot still and the continuous still.

Within the section Maps and Plans of the Codrington Papers, a single page sketch was discovered and labeled "Plans (tracings) of sugar works at Betty's Hope, showing crane, cistern, still room, coolers etc. With plan of proposed arrangement of steam engine and sugar mill by Mirlees and Tait, engineers, Glasgow." This final piece of the puzzle is the key to understanding the integration of steam within the factory works. While there is no clear date for this sketch, it was included with the 1862 schematic of the factory works. The sketch shows a "taiche fireplace" (or boiling train) connected to a chimney of a very similar configuration to that of the western side of the boiling house. A report compiled by architectural historians from the University of Florida in 1994 notes the nineteenth-century addition of a smokestack, which would have been implemented with the steam process. Not visible in their drawings, this edifice is evident to the rear of the boiling house in an old photograph dating to 1910 of the steam-powered sugar works at Betty's Hope (see chapter 7).

The advent of the Industrial Revolution marked a transition from artisanal hand manufacture to the use of machinery and greatly improved efficiency and production yields, often up to 10–15% over previous forms of production (Bradley 2008). British economy had come to rely on the Atlantic trade network and diversity of exports in order to support its economy. The development of steam technology revolutionized production and flooded markets with industrially processed goods. These new scales of production heralded massive modern economic and social growth. Steam replaced huge labor forces, allowing production to continue with much less human interaction. This lowered the need for labor and aided in the movement toward the emancipation of slaves, which was finally written into law in 1833. The oncoming emancipation of enslaved Afro-Antiguans provided landowners a direct enticement to update to steam machinery.

Archaeological Excavations of the Still House

Methodology

Archaeological fieldwork at the Betty's Hope still house site (BH-B) was undertaken during the 2013–2015 field seasons, with the excavations at the BH-B site focusing on structures to the south of the main still house (figure 12.4). Because no historical plan of the factory existed at the time of excavation, the excavations concentrated on understanding how this area of the site was utilized during the production of rum in the eighteenth and nineteenth centuries. During the excavations, we exposed a number of deep stone walls that had been covered with fill. During the 2014 and 2015 field seasons, we followed wall structures up the hill to the east and encountered more walls and more fill deposits.

Excavation units were laid out facing north/south to follow visible features. Trenches were of varying size, some 1 × 2 m, others 1.5 × 1.5 m. Each context was designated a specific number, which was recorded and included in a Harris matrix. The deposit was excavated using a single context system following the stratigraphic matrices of the site. This was often an issue because much of the structure itself was filled with a single-fill stratum. Often, it was only through careful analysis of the material

Figure 12.4. Betty's Hope Still House excavation area.

culture styles within the fill that a stratigraphic matrix could be identified. Because of the structural compartmentalization of the site, locus numbers were designated to mark out internal fill between wall compartments. The stratigraphic profiles that are recognizable do indicate distinctive changes in soil color, which could indicate different depositional events, even from fill. A system was developed to signify areas of identifiable specific depositional events, for example Locus 2(a), Locus 2(b), etc.

Scale plans of each context were completed, mapping in all features, including walls, and each level was documented using digital photography utilizing a photo board. The soil from each context was sifted separately using a 5 mm mesh in a rocking screen. Levels were taken using a total station and mapped into the overall site by Cory Look. Floatation samples were taken and processed in 2013 by Geneviève Godbout; however, very little was uncovered using this method, and it was abandoned in subsequent field seasons.

The finds from each context were bagged and labeled by context, noting the site code (BH-B). All artifacts retrieved as part of the project were then cleaned, catalogued, and curated per the Betty's Hope Project's guidelines before being divided into material class (tobacco pipes, bottle glass, ceramics, etc.) for detailed analyses and identification. Artifacts were numbered, described, weighed and measured, photographed, and entered into the project database. The artifacts were placed into 4 mm plastic bags, organized and packed away into protective boxes and stored at the Field Research Station at Shirley Heights, accessible to researchers by permission of Reginald Murphy.

Archaeological Evidence

After three field seasons, the general consensus was that this is a complex area to excavate and interpret. A number of well-constructed external limestone walls and floors, as well as poorly constructed internal rubble walls, were discovered. Within these features, the context was primarily that of a limestone dust fill interspersed with artifacts that included domestic ware, bottle glass, clay tobacco pipes, buttons, barrel hoops, and diagnostic and non-diagnostic iron machinery parts and lead piping and caulking. A section of lead piping was recovered during excavations (figure 12.5).

In terms of interpretation, it is likely that the most easterly section of the excavated area is a remnant of the "Cast Iron Cistern. 8 feet deep" marked on the 1862 schematic of the factory works. The western area, due to the presence of plaster-lined internal walls and gullies cut into various parts of the new chambers, could be interpreted as an augmented water storage zone, similar to that found at the Galways plantation in Montserrat. An underground channel, opening in the southern wall and disappearing toward the south, may have provided a water source by linking to the nearby pond. This therefore represents the response to an increased need for water driven by the introduction of steam technologies on the plantation.

Dating these changes at the still house structure has proven difficult. Artifact analyses from this area of the plantation suggest a date range from 1780 to the mid-twentieth century. This would correlate with the idea that the current still house dates to c.1780 and that subsequent additions and improvements were accommodated in and around this structure as production methodologies changed to meet market demands for rum and sugar. Shovel-test pits in 2013 immediately to the west of the main excavation yielded no sign of the pre-1780 industrial sugar and rum works, although they may still exist, possibly below the structure.

The broad date range for the artifact assemblage is reflected in the nature of fill found in this area during excavations. This has made standard

Figure 12.5. Lead piping discovered during the Still House excavations.

relative dating through such artifacts as clay tobacco pipes inconclusive for dating structural changes at the still house, even though some clay tobacco pipes are easily recognized, such as the nineteenth-century Turk's-head pipe recovered from excavations in 2014. The sheer quantity of domestic materials found during the still house excavations suggests that perhaps fill was taken not only from the factory works, but also from the nineteenth-century occupation of the manager's house.

The artifacts associated with the still house excavations were categorized by functional type. Most of the artifacts are either production related (materials that came from the act of producing rum), consumption related (from consumption at the still house or elsewhere on the plantation), or structure related (from the building), although there is a small proportion of finds that relate to distribution (the movement of rum), domestic use (materials from within domestic spaces), martial applications (weaponry); and personal use (items of personal adornment such as buttons). The assemblage catalogue was divided during the post-excavation process into these categories and designated by taphonomy to allow for further analysis.

A significant portion of the assemblage designated as production reflects the industrial period. These artifacts consist of tools, nuts and bolts, and small and large pieces of iron from steam machinery. The Codrington Papers note the importation of Bristol iron to Betty's Hope (NAAB Codrington Papers:C17), presumably to be worked by the blacksmith on site.

A large amount of burned material discovered at the site is possibly remnants from the steam engine furnace. Tiny pieces of coal were also found; much of the coal at Betty's Hope was imported and bought from the Jeffersons of Whitehaven, England. Interestingly, an early form of rubber valve was also discovered at the still house. India rubber was introduced to Europe from the East Indies in the eighteenth century and was often used as part of steam engines to seal pipes and make them airtight.

Lead and copper were also found. Lead, now known to be toxic, was commonly used as a caulking material and would have been used to waterproof the shingling on the roof of the still house building for the stills and the pipes used for the distilling apparatus. Worm coils were often made of lead or pewter as were the pipes that carried the liquor between the worm coil and the canning/barreling room at Betty's Hope. Copper, used architecturally, provided corrosion resistance and waterproofing. The copper could also have been part of a copper still, which was usually hammered thin to increase conductivity.

Although we were not surprised by the high number of items relating to the structural and production history of the site, we had not anticipated the high number of consumptive artifacts at a productive site such as this. Excavations at a blacksmith shop (Praetzellis et al. 1997), the New Jersey Central Railroad (Veit and Schopp 1999), and the industrial town of Harpers Ferry (Shackel 2014) have hinted at large-scale consumption (alcoholic and otherwise) in workspaces. However, in these areas, either the material culture being used for the work was dramatically different from personal consumption vessels or there was a definite difference between the bottles being used for work and those for personal consumption. In the case of the still house at Betty's Hope, there was no way to separate vessels being utilized for production from those being used by laborers.

While much of the rum was canned (in crock jars, for example), some would have been put into barrels (as would the majority of the sugar). Several barrel hoops were found on the site; barrel hoops were often shipped to Betty's Hope from Britain through the Codringtons' many merchant contacts (NAAB Codrington Papers:A68).

Discussion and Conclusion

The excavations of the still house at Betty's Hope reflect a dynamic and changing landscape in rum production over several centuries. The material culture of an industrialized site like this offers interesting possibilities for understanding the materiality of rum production on a Caribbean plantation. The act of mass production allows the archaeologist to examine the steps taken by the producer to produce the object, which, while highly utilitarian in nature, can also be construed as a "production of commodities [that] is a cultural and cognitive process" (Kopytoff 1986:64). The industrialization of the factory complex generated a newly habituated space, changing the workspace and reflecting many economic and social changes.

The modern landscape of Betty's Hope still displays this industrial heritage. The remains of sugar cultivation tools and industrial machinery lie discarded and abandoned on the property. The steam locomotive and engine, as well as boilers, are principal signifiers of British production in the nineteenth century, retaining an almost iconographic place in British Caribbean history. More specifically, the development of rum as a material culture, a material with which humans develop a direct, intensified relationship, caused the modification and adaptation of mentality, methodology, and use of space at the Betty's Hope still house and factory complex. Modifications within the Caribbean factory affected the development of consumption styles and methods, driving demand, facilitating the formation of the consumer revolution, and contributing to consumer culture. Rum, along with other vital commodities of the period, left an indelible financial and social mark on the global world, forming penetrating, assorted new market systems and social hierarchies.

13

Beyond the Plantation

The Codringtons, Betty's Hope, and the Defense of Antigua, 1670–1714

CHRISTOPHER K. WATERS

Between the arrival of Christopher Codrington II on Antigua in 1674 and the emigration by his son Christopher Codrington III back to Barbados in 1707, Britain was officially at war for 21 years. Internecine warfare made every Caribbean island a target for raids and conquest, encouraged by the nascent and growing sugar wealth driving industry and national economies back in Europe. Thousands of kilometers from Europe and—given their small size—with nowhere to retreat to, the islands of the Lesser Antilles developed their own strategies of defense to cope with the very real threat of invasion, with each island focusing on its strengths and mitigating its weaknesses. Each island experienced defense in its own way according to its physical location, natural features, nationality, demographics, and experiences of the plantation elite (Leech 2013; Machling 2012; Verrand 2004; Waters 2017, 2018, 2019).

Defense and the projection of military power are at the core of an academic understanding of colonialism, rooted in both the colonizers' respective histories and competition over economic and cultural resources and manifested on the landscape (Dietler 2010:27–43; Given 2004; Gosden 2005). Because of the islands' small sizes and their physical distance from the metropole, geographic barriers between islands created strict boundaries and forced together people from different European and African cultures (for a French comparison, see Dawdy 2008). These factors contributed to each island fostering a strong sense of local independence by the plantation elite, resulting in a plurality of different approaches to the creation of a colony. Mark Hauser (2011:122) comments that in Jamaica, "there was in fact a tremendous diversity of social, cultural, economic practices in colonial Jamaica, and colonial/postcolonial models . . . only capture a part of that diversity." This statement is as true for Antigua as it is for the rest of the islands in the eastern Caribbean. Nowhere is this seen historically or archaeologically better than in crafting defense policies by the local colonial governments.

The Codringtons and Antigua's Defense Policy

Two major political changes occurred following the successful French invasion of Antigua in 1666/1667. The first was Antigua and the rest of the Leeward Islands successfully sued for their political independence from Barbados, which, they contended, refused to send supplies or reinforcements. This led to considerable local self-rule by Antigua's elite (Dyde 2000; Waters 2018, 2019; Zacek 2010). The second was an acknowledgment that the island was extremely vulnerable to attack, making a defense network of guard houses and fortifications necessary to prevent further attacks. The economic losses from the French attack devastated Antigua, and despite being the largest of the Leeward Islands, the economic repercussions from the attack convinced the local elites that it was in their best interest to invest in island-wide defenses to protect their property.

In assessing Antigua's defensive capabilities, a military engineer sent from London commented that "the Coast is indented with Bays, the approaches to which are rendered difficult by numerous Rocks and Shoals" (Lancey 1831). Compared to its volcanic neighbors, Antigua is relatively flat, with a large number of bays and inlets punctuated by hundreds of sandy beaches promoting the rapid movement of people by water but also posing a serious hazard for an attack. The sheer length of the coastline made defense much more difficult than the long, regular coastlines and cliff faces, natural features that limited the number of places at which a landing could take place. This vulnerability is reflected in the 1703 Act Empowering a Commissioner to "Sett Forth our weak and Hazardous Circumstance in this Dangerous time of Warr," referencing:

> Whereas Since ye breaking out of this present Warr, wee have been and Still are in Continuall danger by the great Number of Privateers daily Insulting our Coast, Landing in our Creeks and private Inletts, Takeing off our Slaves and Committing other frequent Depredations

> to ye Continuall harassing the few Inhabitants yett remaining who are greatly discouraged for want of Sufficient Force to leave their Plantations from the like Injurys for the Future. (1668–1967, NACO 8/3:143)

Engaging privateers under special commission issued by a European power was a common war tactic to enrich private individuals who were willing to outfit private vessels. Privateers also functioned as part of a larger strategic interest in destroying the economy of an adversary at little cost or risk by the granting country (Elleman and Paine 2013; Marx 1992; Rediker 2004; Zahedieh 1986). Although privateering is generally considered a water-based institution, the small Caribbean islands, often with large stretches of unguarded coastline, offered a secondary source of income through raids (Zahedieh 1986). Antigua in particular suffered early on from "The Frontiers, and other parts of the Island Lying naked, and Exposed to the frequent Incursions of the Enemy for want of Standing Guards" (July 27, 1704, NACO Antigua Sessional Papers:8). This not only caused individual material loss, but during times of war, also restricted supplies of food, equipment, and labor on which Antigua depended for survival. One solution was to outfit their own privateers and rely only occasionally on Royal Navy vessels, which arrived in the Lesser Antilles in small numbers and infrequently, despite numerous pleas to the Admiralty. A more proactive response was the creation of a fortification network to cover the entire coastline.

Christopher Codrington's arrival in Antigua in 1674 and his economic and political success mirrors Antigua's own evolution from a small staple and cash crop economy to a sugar monoculture (chapters 2, 6, and 7 in this volume; Sheridan 1989:60). Antigua and the rest of the Leeward Islands were amalgamated in 1670 into their own federated colony of four neighboring islands, allowing them to confront their own problems through local self-government. One of the key hurdles for Antigua was overcoming its small population, which made it economically marginalized despite its being the largest of the Leeward Islands with the most arable land (Dunn 1972; Engerman 1996:155–159; Sheridan 1974; Wells 1975:126–146; Zacek 2010).

In 1678, Antigua's population was 4,480, with almost half of the people comprising enslaved Africans (Dunn 1972:141; Wells 1975:209–212; Zacek 2010:48). By the time of Christopher Codrington III's departure from Antigua in 1707, the white population grew only slightly, to 2,892, whereas the enslaved population burgeoned to 12,892, comprising almost 82% of the entire population (Dunn 1972:141; Southey 1827:200). By 1707, over half of the Leeward Islands population, scattered over four large islands and several smaller dependencies, lived in Antigua. The population growth in Antigua reflects what Engerman (1996:158–159) sees as a rapid need to maintain regional sugar exports as Barbados's sugar exports dropped precipitously as a result of depleted soils at the end of the eighteenth century, and Jamaica's economy had not yet switched over to large-scale sugar cultivation.

Under the Codringtons, Antigua grew increasingly valuable economically, as sugar cultivation took off, and militarily, due to its position just north of Guadeloupe and east of the rest of the Lesser Antilles. With the winds, a ship from Antigua could rapidly reach these other islands for a surprise attack while also dispatching reinforcements to the other British islands. When war broke out in 1689 and the French gained control of St. Kitts, Governor Stapleton resigned his position, elevating Christopher Codrington II to governor and captain-general of the Leeward Islands. Outgunned and outmanned in the initial stages of the war, Codrington managed to instill confidence in the English defenders during the early years (Parker 2011:83).

Mustering volunteers and militia and reinforced with regular-army troops and Royal Navy ships sent from England, Codrington led a successful campaign against the French, recapturing St. Kitts in 1690. His success, however, was limited. While he did organize raids on other islands—under the guise of attacking military targets but quite likely as enriching plundering expeditions—outright military success seems to have eluded him. Guadeloupe was attacked in 1691, pushing the French defenders into their fortifications high on the mountains. The invasion was foiled only by the arrival of a French relief fleet, which forced the English to retreat. An attempt on Martinique was made in 1693 with disastrous consequences, the English losing over 1,000 men, mostly to disease. It was in this latter campaign that his son, Christopher Codrington III, an Oxford graduate and successful veteran of battles in Flanders, was reunited with his father after more than a decade and returned with him to Antigua.

Major military actions were largely circumscribed until the end of the war in 1697. However, raiding on Antigua and the other Lesser Antilles continued to take place as Codrington struggled to maintain morale and the defense of the islands under his command. Effective French privateering prevented supplies from reaching the Caribbean. The will for further military action, especially carried on the shoulders of local militiamen from the islands, broke. Christopher Codrington II's last years before his death in 1698 were spent in a whirlwind tour of the Leeward Islands, attempting to bolster defenses and persuading the local governments to continue to pay for their own fortification networks (Parker 2011:180–191; Webb 2013:277). Although he had limited success on the remaining Leeward Islands, his ideas percolated in Antigua, pushing to protect the weak, underpopulated east coast of Antigua. Besides continued contributions to the citadel at Monk's Hill, he commissioned a network of guard houses built along the coastline, with regular detachments of armed militia patrolling in between these guard houses, regularizing the need for defense among Antigua's elite (NAAB 1693–1747:314:84; see also Buisserett 1973).

Fortifying strategic points along the coastline provided both deterrence and an active defense against raids and attacks. By 1676, Governor Stapleton made reference to a third fortification at Carlisle Bay, with Falmouth and St. John's each already having a formal defensive works in place (Stapleton as cited in Oliver 1894–1899:III:liii). This was followed by the dramatic action to fortify the top of Monk's Hill into a deodand, or place of last retreat. With an explicit, frank acknowledgment about Antigua's small population, the construction of the fortification is enshrined in Act No. 138, where it states in the preamble that the purpose of the fortification is "given by the frequent Insults of our Enemies; and as Nothing, in case of an Invasion, can more encourage small Numbers of Men valiantly to behave themselves, than the Consideration of having their Wives and Children secured" (Buisseret 1973:51–55; Legislature of Antigua 1805:175; Nicholson 1994:6–7; Oliver 1894–1899:III:lxix).

By 1701, the remaining Antiguan trading towns at Bridgetown and Parham registered fortification, although Parham did not have any mounted cannon (Oliver 1894–1899:III:lxxi). With the beginning of the Queen Anne's War (1702–1713), defense again became a major priority, with a list of 24 fortified sites along with a list of officers who were appointed to take charge of the fortifications (NACO Antigua Sessional Papers:24–25). This list forms the core of the fortification network that would guard Antigua for the remainder of the eighteenth century.

The Forts of Antigua

In Antigua, the fortification network, which the Codringtons advocated for, was a long network of guard houses, platforms, and gun batteries, anchored by the single large citadel at Monk's Hill. Whereas Monk's Hill was the closest representation to a European-style artillery fortification on Antigua during the early years of the colony, the remaining fortifications were small defensive points scattered along the coastline. Although Christopher Codrington III was deposed as captain-general at the end of 1703 for many failures, among them a disastrous campaign to take Guadeloupe, it is nevertheless possible to gauge the defensive legacies of his fortification plans in Antigua.

In September 1704, the Antigua Assembly commissioned a report on the state of Antigua's fortifications, which lists 23 completed defensive sites around the island and an additional four platforms under construction (NACO Antigua Sessional Papers:24–25). Their positions highlight both the underlying strategy and the weakness of the system: protection for plantations and busy harbors, rather than a systematic defense of the entire island (figure 13.1).

While the defensive potential of these fortifications is generally in question, their placement on the landscape demonstrates a significant adaptation of a colonial ideal in the seventeenth and early eighteenth centuries: protection of wealth and the wealthy. Antigua's central plain cutting through the middle of the island is the most fertile area and the area of the earliest plantation developments, whereas the more mountainous south seems to have been developed much later. Likewise, the central areas offered easy access to the ocean, a key component for trans-island shipping of heavy hogsheads of sugar to the customs houses in one of the official trading towns. With the heavy reef systems along the northern coast, only shallow-draft vessels and local knowledge of the channels created a relatively protected coastline. This coastline, then, was protected only by a series of guard houses, permanent small structures where a small group of men could keep watch for raids (NACO Antigua Sessional Papers:8). Two of these structures still exist archaeologically: Thomas Bay and Pearne's Point on the west coast. Both of these structures consist of a single rectangular stone house of approximately 8 m^2. The guard house at Thomas Bay also has a small stone platform, most likely for a single signal cannon. Both of these sites are within visual range of the next fortifications along

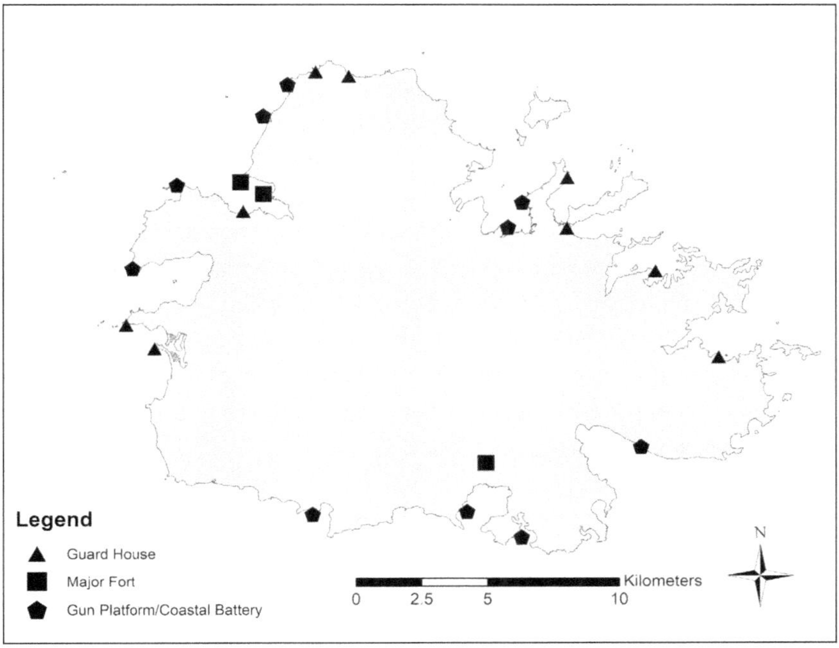

Figure 13.1. State of Antigua's fortifications in 1704.

the coastline, situated on promontories. The purpose of these defensive structures bears out in the large number of alarms recorded in the documentary record.

Betty's Hope and the adjoining Codrington Cotton Plantation stretch from the central plain northward to the coastline, following the waterline. The 1710 map of the plantations includes a small image of a fortified storehouse. This is a stylized image, with the multistoried crenellated structure drawn to look like a blockhouse (for examples, see Coe 2006; Hart 2013). The original guard houses were all built by, and under the command of, local plantation elites and their expenditure compensated for by the Antiguan government (NAAB 1693–1747:314:89, 94). This public/private enterprise suggests that a large variation in size and construction is possible, with each individual planter choosing to construct what would best suit him.

The Cotton Plantation guard house appears in the 1704 return of fortifications as one of the official defenses of the island. The expansion appearing on the Cotton estate map suggests that the Codringtons actively enhanced the public space, expanding it for their own purposes of protecting their crops. Location, as well as personal wealth, played a role in the deviation from the small guard houses at Hawksbill and Southpoint; the latter two are built on high bluffs, whereas the Cotton Plantation guard house is along the water's edge. Exploiting their ownership of the land, the Codringtons as private citizens were able to exploit the infrastructure project proposed by themselves for their own gain. Indeed, this is not the only case in which individual planters exerted control over the fortification planning; 12 of the 23 fortifications listed on the 1704 return are named after plantation owners. This convention continued throughout the eighteenth century, where each fortification was unique, conforming to the will of the landowner rather than principles of fortification (for the wide variation, see Horneck 1752).

While the individual plantation owners exercised control over the individual defensive points on their properties, geography and a consensus of important strategic considerations by the Antiguan government dictated where these fortifications were located on the landscape. In that regard, the main concern centered on protecting shipping. The west coast of the island demonstrates a large number of fortifications regularly spaced along the coastline. The principal harbor and town, St. John's, is situated along this coast. Based on shipping returns recorded between 1704 and 1720, the majority of the shipping cleared through the customs house in St. John's (NACO Miscellanea:157/1).

As St. Johns' was the most populous town and a hub for trade, its harbor was highly important to the plantocracy. Defensively, however, the harbor provided both a boon and a challenge. The harbor itself was obstructed by a sandbar running across the entrance, with depths as shallow as 12 ft. and somewhat less at low tide (NACO Maps and Plans). John Luffman (1789:25) describes this sandbar as St. John's Harbour's "best security," despite the large number of fortifications surrounding the harbor's entrance. This seriously precluded the size of vessels entering St. John's Harbour, forcing larger vessels to ride off shore, where they could anchor safely (Baker 1749). In his classic *Architectura Navalis Mercatoria*, Chapman (2006 [1768]:vi) suggests that most if not all ship-rigged vessels would have a draft deeper than 12 ft. For example, a West Indiaman was recorded by Chapman as having a draft of 15.75 ft.

Of the 37 vessels loading sugar for export between May 27 and September 15, 1704, 22 were ship-rigged and were unlikely able to make it across the sandbar (NACO Miscellanea:2). The larger ships necessarily had to anchor outside of the harbor. Here they were vulnerable to attacks by privateers looking to cut ships away. For instance, the ship *Hannover,* delivering a cargo of sundry dry goods and merchandise to Antigua in 1704, was "unhappily taken out of the Road at St. John's by a French man of Warr," right under the chain of coastal fortifications along the west coast (NACO Antigua Sessional Papers:26). Access to the Atlantic World only came through allowing ocean-going vessels safe access to Antigua.

Interdiction and blockade not only cut off an avenue to ship sugar off the island—the source of most of the island's wealth—but also prevented supplies from coming in. As a small island, Antigua had limited raw materials and relied on imports of everything from food supplies to lumber to keep its economy running. For Antigua's government, the immediate need to protect its most vulnerable coastline to shipping interdiction was a consensus item, and there was general agreement about the importance of creating a concentration of "the Great Guns along ye Bay between the Fort & Dixsons Bay" (NAAB 1693–1747:316:24) to protect St. John's Road.

The south and east coasts of Antigua, the most vulnerable to attack, were considered strategically less valuable by the plantation elite. With the prevailing trade winds blowing from the east to the southeast, Antigua was the first landfall for ships making their way across the Atlantic to the Caribbean. The south coast faces the French island of Guadeloupe, and although Antigua lies slightly to windward and is thus somewhat more difficult to reach by a sailing vessel, the coastline was still approachable. While not protected by reef systems and shallow water like the north and parts of the west coast, the east coast does have limited reef protections, especially around Willoughby Bay and the seventeenth-century trading town of Bridgetown.

The south coast is covered in steep mountains with few beaches. The water is rough and deep here too, with the Atlantic storms driving the swell, allowing vessels to come in close to the coastline. The fortifications built by 1704 reflect the tactical considerations of these coastlines with only the port towns receiving any kind of protection. Old Road, Falmouth Town, and Bridgetown were declared official trading ports, and each of them fronts onto a sheltered bay. The immediate hinterland, however, is not ideally situated for large-scale sugar plantations.

Although a few plantations did exist, most were very small and given over to other cash crops, especially cotton, which grew in more arid conditions. The purpose of these fortifications was immediate protection of shipping and the town, ideally against privateering rather than a full-scale assault. Some of this shortcoming was already recognized by Antigua's elite in 1704, when they proposed a further expansion of four platforms, two of which would augment the defenses of Falmouth and Bridgetown with new fortifications of three cannon each (NACO Antigua Sessional Papers:25). But with only two representatives from this side of the island in the Antiguan government and the majority concentrated around St. John's and the north coast, this area received little attention. These expansion sites did not remain a priority; the secondary fortification for Bridgetown was not built until March 1739, 35 years later (March 14, 1739, NAAB 1693–1747:324).

The early fortification network was based on tactical considerations, predicated on protecting the wealth-generating parts of Antigua rather than a holistic, strategic plan to make the island impregnable. Caught between their desire for prosperity and burdened with the protection of the island, the Codrington governors plotted a defensive strategy that relied on the militia's throwing an invasion force off the island as they landed rather than on creating powerful defensive structures. This initial project focused on a public/private venture, where the wealthy were tasked with creating the fortifications to their own specifications and reimbursed from the public treasury for their time and effort in constructing and commanding these defensive structures.

The Cost of Fortification

As a local government representing the needs and desires of the elites, the fortification network can easily be interpreted as a desire to protect shipping, and thus wealth, the purpose of the colonial enterprise. If, however, the plantation is the nexus for wealth generation, then the defense system must be seen as a drain on the local economy. Beyond the necessary labor and pay requirements, the local government was constantly enacting occasional taxes to support the public debts of the island as needed. Regular revenue came from fees collected for official documentation and registration, an excise tax on liquor, and a general import tax. However, this revenue stream seems never to have been enough, and additional monies were constantly needed. Special taxes were passed at least nine times between 1668 and 1706, raising tens of thousands of Antigua pounds and hundreds of thousands of pounds of sugar for the settlement of public debts. An analysis of the Council Sessionals from Antigua between 1692 and 1704, which recorded the money spent, shows a large proportion related to defense, including pensions for injuries and loss of life by militiamen, construction and maintenance of fortifications, bounties given to privateers, and compensation to individuals for providing loans of money and goods for the purposes of defense. Given the small tax base on the island, the majority of the money came from the plantation elite. A legislative act raised a special tax of 1,240,000 pounds of sugar in 1697 levy rates on acreage and the number of enslaved persons and included a special excise on existing stocks of liquor, a commodity that was already taxed at import (NACO Acts of Antigua 8/3:5). While this tax does target freeholders, there is a disproportionate financial burden on those with land and enslaved labor.

The fortification system carried a heavy financial cost for the island's elite, clearly reflected in their rapid ratcheting up of military spending during times of war and just as rapid demilitarization and neglect. The social cost of the fortification system, however, was borne out by the poor white settlers who were heavily affected by their legal requirement to provide money and labor for defense. A commentary in 1703 suggests that

> Antigua in particular is extremely weak in its inhabitants, the number of whom is considerably lessened, and those few are so greatly fatigued by guarding, by frequent allarms and marches for the necessary defense of the coast, that they apprehend themselves in danger of losing all their settlers of the poorer sort, who cant support themselves and their families under the frequent and expensive attendances abroad to which the weakness and nakedness of the Island obliges them. (NACO Miscellanea 157/5:39; 153:8. pp. 215–219)

This passage, taken from a letter from Lieutenant Colonel Thomas to the Council of Trade and Plantations, succinctly states the limitations faced by Antigua's defenders: there were not enough of them to adequately provide protection, and the burden of protection fell on the poor white settlers and indentured servants, forcing them to neglect their own welfare. The result was that Antigua was never an attractive place to settle, pushing away those who did arrive. This same phenomenon is commented on again during the Prince Klass Slave Conspiracy in 1736, during which the militia was called out for months at a time, causing friction between the poor whites who filled the ranks and the plantation elite who were more or less exempt from the daily guard activities (Gaspar 1985:21).

The economic detriment to the island of removing small landholders, traders, servants, and plantation workers from their duties would have been significant. The social impact, however, is reflected in the demographics of Antigua, where the white population peaked in 1724, the same year that all available land on the island was patented, with the vast majority tied up by 300 plantations (Gaspar 1985:68; Oliver 1894–1899:III:xcv). Furthermore, the land that was apportioned to these white families was along the south and east coasts, the areas where the likelihood of an attack was higher.

As more and more poor white farmers lost their crops and livelihoods to guard and militia duty, the more indebted these small holders became, forcing many to sell out and abandon the land, which continued to affect the demographics of the island. The entire defense system was predicated on being able to attract enough of these men to Antigua and provide them with a subsistence-level living. When these men failed to stay, the whole system was placed in jeopardy. Despite the best efforts by the Codringtons and their successors to attract more settlers, they were constantly stymied by a lack of land and the constant need to expand landholdings of the elite, including their own holdings. As the eighteenth century wore on, the shrinking population was burdened with more fortifications, imposing even larger burdens on the smaller white population and creating a feedback loop that persisted until the disbandment of the Antigua Militia in 1838.

Conclusion

Antigua's early fortification system is as much a product of the nascent plantation landscape around 1700 as it is of the flexibility practiced by a group of elites controlling the political, military, and social systems and making it work for their own needs. While successful insofar as the island was never invaded after 1666 (unlike Antigua's neighbors), Antigua's early defenses are more a study in the creation of a colonial system and the establishment of a plantocracy than a model defensive network.

The military in colonial contexts often gets swept up in an imperial narrative. In Antigua's case, and indeed in the eastern Caribbean, the fortification system was a local reaction to its role and position within the Atlantic World. The dictates did not come from a perfectly conceived metropolitan-colonial ideal, executed with the support of national coffers. Rather, the fortifications of Antigua were an outgrowth of a colonial system that was being enacted and thought out locally as time passed, reliant on the needs of the plantation elites.

Antigua's defense network of fortifications during the early years of the colony was borne out of a region wracked with war. The islanders were well aware of the financial cost of facing an invasion, both suffering under attacks and participating in attacks on neighboring islands. The Codringtons, as wealthy plantation owners and as the heads of the government of Antigua, played a major role in crafting the defense network surrounding the island. As private individuals tasked with building and maintaining a guard house on their own property, they manipulated their public mandate to benefit themselves, extending a guard house to a fortified warehouse at the public's expense.

As public officials, the Codringtons led invasions against their French neighbors and fended off raids by European and indigenous enemies seeking to attack Antigua. As military leaders, the Codringtons understood that even the most well-built fortifications in the Caribbean were still vulnerable. Thus, they focused on creating a chain of fortifications to protect the most vulnerable parts of Antigua's infrastructure: shipping. As safe shipping was necessary for Antigua's elites to generate more wealth and income, the fortification project focused on those areas with the largest number of plantations and on the main ports on the island.

The coastlines settled primarily by poor white farmers on 10-acre plots were protected only occasionally, if at all. A direct consequence of this policy was the continual emigration of poor whites from the island and an expansion of the fortification system throughout the eighteenth century. Ultimately, the fortification system caused Antigua financial ruin, ending its ability to self-rule with little political interference from the metropolitan government.

PART VI

Cultural Heritage Management at Betty's Hope

14

The Restoration of the Betty's Hope North Windmill, Antigua, West Indies

REGINALD MURPHY

This chapter examines the technology, archaeology, and restoration of the extant and iconic north windmill at Betty's Hope, located to the west of the former great house. Of the two windmills at Betty's Hope, this earlier windmill epitomizes the large-scale operations of an agro-industrial sugar complex and the role it played in the processing of sugarcane. The main windmill restoration project was conducted from 1988 to 1995 as part of the cultural heritage management and initial archaeological investigations of the site. Restorations of the north windmill continue, with the most recent replacement of a new tail pole by Jerry Bardoe in 2015. This chapter will cover only the archaeological investigations of the windmill and main restoration project conducted between 1988 and 1995.

Windmill Technology

The windmill tower was a conical-shaped structure built of stone. On average, the walls were 1.2 m thick at the base, decreasing or tapering to about 1 m at the top. The early mills were roughly 16 m high, generally with three openings. These comprise two large arched doorways on the leeward side through which the cane was brought into and taken out of the mill; a tall narrow exchange slit for repair work on the shafts; and a small opening at ground level for the cane juice to flow through. A small arched fireplace-style opening was placed about 2.1 m above the ground into which lanterns were placed for night operations.

The type of mill used depended entirely on the available resources of each plantation site. On islands with rivers and running streams, water-mills were the preferred option, but as Antigua has no rivers, wind or animal mills were the only options. By the late eighteenth century, there were 175 windmills in operation on Antigua, generally one per estate. Wind-powered mills were built on the highest point of land on the estate in order to catch the most favorable and unobstructed flow of wind. The higher elevation also facilitated the flow of the cane juice to the boiling house. Some of the largest, wealthiest estates had two windmills, and Betty's Hope was one of them.

Today, these stone towers, about 90 of them, are all that remain of the windmills scattered across Antigua. They have been recognized as a prominent part of Antigua's cultural heritage and can be observed on the national crest of the island, which features a windmill tower flanked by two fallow deer. Wind-powered mills on Antigua would remain in use until the twentieth century, but the efficiency of steam as an energy source was adapted by the 1860s (Lowes 1994b:18; see chapter 7 in this volume).

As a large and lucrative estate, Betty's Hope could afford to adopt the latest steam-powered technology as the energy source to drive the mill rollers. According to the Codrington Papers, a proposal to upgrade the steam engines was submitted in 1862, and by the 1880s, Betty's Hope was dedicated to steam power (see chapter 7). The milling operation had been moved to the north end of the boiling house when the first steam engine was installed and the windmills abandoned. The openings in the south mill were filled in, and the stone tower was converted to a water tank that was filled by a wind turbine water pump (fan mill) from a nearby well. By 1878, 35 of the major sugar producers had already changed to steam, and by 1894, 59 of the 71 sugar estates in operation had made the change (see Lowes 1994b:18).

The early 1920s saw the establishment of a central sugar factory complex and the syndication of most of the sugar estates into a single production and management authority of which Betty's Hope was a part. The last wind-powered mill in operation on Antigua was finally abandoned in the 1920s. Over time, all of the sugar mills on Antigua were plundered for their valuable cast-iron machinery and bronze bearings. Some were salvaged for export back to Britain in support of efforts for both world wars, but many were stripped by scrap iron merchants and sold.

At the time of the restoration of the Betty's Hope windmill, only three complete sets of cane-processing rollers remained on Antigua. Many of the iron rollers found new life as rollers for compacting cricket pitches at all cricket fields on Antigua, and many are still in use. In comparison, very few of the massive iron cogs, spindles, and shafts that comprised the internal mechanisms that drove the rollers were left on Antigua.

Considerations in the Restoration of the Betty's Hope Windmill

In 1988, a committee of the Historical and Archaeology Society of Antigua and Barbuda decided to restore a windmill to working condition. Previous archaeological surveys of the surviving mills and abandoned sugar works on the island had located and mapped all the parts necessary to do so. The windmill chosen for restoration was Betty's Hope's, and the Betty's Hope Restoration Committee was formed. The Betty's Hope estate was selected for a number of reasons. These included government ownership of the property; the enormous archaeological potential of the site (50 acres); the relatively good condition of the site's twin windmill towers; the site's major historical importance for Antigua and Barbuda due to its connection to the Codrington family and the extensive Codrington archive (see chapter 2); the location, which was ideal for the development of the site as a national heritage park and as a tourism attraction; and the importance of windmills and Betty's Hope in the cultural heritage of Antigua and Barbuda.

The decision was made to rebuild a windmill typical of the late eighteenth century, as the available machinery on island dated to that period. All mills used three rollers through which the cane stalks were fed. The earliest mills of the late seventeenth and early eighteenth centuries used a system by which the rollers were placed upright or vertically in a straight line. All were connected with iron teeth or geared collars/cogs at the top, with the center roller directly attached to the main vertical driveshaft or spindle. When the center roller was turned, the connected cogs turned the outer rollers. These early mills required two operators, one on each side of the machine, feeding the cane stalks into the rollers and back for a double crush, a time-consuming and inefficient process.

Major improvements were made to the system to reduce production costs and extract a higher yield (see chapter 7). The most important development was the positioning of the three rollers horizontally, two on the bottom and the third on top. This system reduced the number of workers

Figure 14.1. Photo of north mill. Note the narrow opening of the exchange slit.

required as the cane stalks received a double crush in a single pass and greater pressure could be applied, which extracted more juice from the cane.

Mill towers were also improved, and many were enlarged and raised in height to over 9 m on average. Adjustments were made to the arches in the tower walls to improve the efficiency of the work process, such as new openings for bringing in and removal of the cane and extending the tall narrow open slit (the exchange slit), which permitted the removal and placement of the longer driveshafts (figure 14.1).

The late eighteenth-century mills on Antigua had three floors or levels internally. The first was the ground floor, which was a solid stone-work platform where the rollers were positioned and the cane stalks brought in, crushed, and removed. Above this floor were two others that were made of wood. These upper floors permitted access to and supported the large internal shafts and cogs that turned the rollers. The top floor, or cap house, contained the massive timbers and iron cogs/gears that connect the crown wheel atop the central shaft to the spindle or wind shaft. The wind vanes turned the top spindle shaft that had a small cog that sat on top of the large crown wheel. The smaller cogs then turned the crown wheel, which turned the central shaft and rollers.

The cap house sat on top of a polished wooden ring, normally made of lignum vitae wood, set into the stone masonry. This polished ring had to be kept lubricated to allow the cap house to turn into and out of the wind. Attached to the cap house on the leeward end was a long wooden shaft or pole, or tail tree, which protruded down from the top of the mill angling out about 15 m from the mill wall. This wooden shaft was essentially a lever that was secured firmly to the frame of the cap house. It was supported on the ground by a small iron wheel, generally from cannon. To start or stop the mill, oxen were used to pull the tail tree, which turned the entire upper structure into or out of the wind. On the windward side of the mill were the sails, or points, which was a lattice frame with a wood/plank cutting edge bolted on to the four arms. Canvas sails were spread on the lattice frames according to the amount of wind. The entire operation could be likened to sailing a ship, with the mill operator often referred to as the captain and his assistant as the boatswain.

Archaeological Excavations of the North Windmill

The north mill that was selected for restoration had been rebuilt in 1737 by Richard Buckley. The restoration process began with an archaeological excavation in January 1989. The expectation was for a short, controlled, and expedient excavation that would end with the exposure of a stone floor platform. This was not to be the case, and excavations continued every weekend until August 1990 (figure 14.2). Excavations at this first stage

Figure 14.2. Archaeological excavations at the north windmill.

were conducted by the author, then secretary of the Antigua Archaeological Society and chairman of the Betty's Hope Restoration Committee, as well as volunteers from the society.

Excavations units of 2 m^2 were initially dug in 10 cm levels, with all soil passed through 3 mm mesh screens. This method was changed to a cultural method on realizing that there was no stratigraphy. Units around the perimeter of the interior quickly ended on a lime-mortar rubble stone base 15 cm to 20 cm below the surface. It was soon discovered that the stone floor had been removed. In the center, however, there was an open stone-lined pit about 185 cm deep into which scrap iron from the various works on the estate had been dumped. At the bottom of this pit was a lower floor where the four large iron clamps and open sockets in the stone masonry indicated that there had once been a heavy 30 × 30 cm timber frame and beams onto which the crushing machinery was bolted. On the east side of this large open "sump" were the remains of a large cast-iron tank that was set on a base of bricks. The juice from the mill was collected in the cast-iron tank and then pumped to the boiling house on demand. It was an unusual type of operation that the restoration team had not observed on Antigua or elsewhere. Upon locating the original lower floor, archaeologist Conrad "Mac" Goodwin was commissioned to excavate the undisturbed levels below.

The archaeological findings from the upper levels—the scrap heap—represented an interesting assemblage of industrial artifacts that included cast-iron pipes, iron flanges, furnace doors, metal stencils for marking shipping crates and barrels, skimmers and ladles for the boiling pots, massive hooks and iron rings, foot valve strainers for wells and water pumps, sheets of lead, and much more. In sum, with the introduction of the steam engines and new crushing machinery being relocated at the north end of the boiling house, the north windmill appears to have been abandoned. Portions of the cut stone floor were removed and recycled, and the mill tower with its open pit was used as a storeroom for scrap iron.

The Restoration of the Windmill

The restoration process began with the replacement of the missing sections of the stone floor and repairs to the masonry. The 12 sq. in. hardwood timber beams (*greenheart*) required for the floor beams and shafts were sourced from a mill in Brazil and shipped to Antigua. The crushing machinery was salvaged from Miller's Estate and donated to the project. The spindle, crown wheel, and gears were salvaged from the Judge's Estate mill. All of the machinery was owned by businessman Mr. Ferdinand "Ferdie" Shoul, who donated it to the project on the condition that the project remained in the control of the NGO, Betty's Hope Committee. In 1990, the milling machinery was exported to Whiting Iron Works in Rochester, New York, where the missing bronze bearings were recast, the rollers and frame refurbished, and then returned to Antigua where it was installed at Betty's Hope (figure 14.3).

The key persons involved in the restoration were shipwright Jerry (Chippy) Bardoe (and his team of carpenters); industrial archaeology specialist David Rollinson; Lawson Whiting of Whiting Iron Works; Doug Leury; Tim Degarve, manager of the US Air Station; and project manager Birgit Carstensen. Valuable information was provided early in the project by the only living person we could find who remembered running a windmill and was involved in repairing the mill at Morgan Lewis, Colin Webster of Barbados.

The work was finally completed in 1995, and the first test run was conducted in June of that year by Jerry Bardoe and volunteers, including

Figure 14.3. Refurbished machinery in the mill.

Figure 14.4. Cap house ready for lifting to the top of the mill.

Figure 14.5. Completed windmill with sails, prior to test run.

students attending the Antigua Archaeology Field School from Trent University (figures 14.4 and 14.5). The official opening ceremony and the first public running of the mill was in July of 1995. The project, which depended entirely on volunteers and financial donations, was an outstanding success, and much was learned in the process before the running operations were concluded in 2000. It was never the intent of the Betty's Hope Committee to run the mill continuously, but rather to create a heritage park where the completed windmill could be observed and the process of operation explained in the on-site interpretation center. Antiguans at last had a full understanding of a cultural heritage icon, and a new attraction for both locals and visitors was created.

15

A Scenic Route to Interpretation

The Betty's Hope Visitors Center Exhibition as Cultural Heritage Management

AMANDA KRAMP

Betty's Hope Plantation is important and multifaceted in its historical and social significance related to the production and economy of sugar and to the Codrington family, as demonstrated in this volume. The importance of the history of Betty's Hope directly relates to maintaining the heritage of this site for both locals and the international community. Consequently, as a tourist destination, and given the number of tourists who visit Betty's Hope almost on a daily basis, this has implications for the site's management.

Under the Betty's Hope Trust, the property is maintained by locals; however, oversight is minimal at best. Renovations, chiefly in the restoration of the north windmill and replacement of the outdoor signage and other small improvements on the property, have helped in the maintenance of the site, but weather, wandering domestic livestock, and humans all take their toll. Additionally, the wear and tear on the exhibits in the visitors center called for a concerted effort to update and renew the space. From the fall of 2013 through the summer of 2014, a project was undertaken to apply core concepts and methods from contemporary museum studies and cultural heritage management practices to create and install an interpretive exhibit in the visitors center space. This chapter will focus on the project's work and its results.

Approaches to the Project

Because accessibility is one of the hallmarks of contemporary museum practice, this project was undertaken to work with and consult local Antiguans and to create an accessible, inviting, and cohesive exhibition space. Operating within constraints of an old building and a space that is susceptible to the elements, consulting with both scholars and locals was essential. The project involved multiple layers of interpretive information and the fabrication of exhibit components. The project therefore acts as a case study by emphasizing the practical application of theoretical perspectives and enumerating the process of exhibit development and installation in a developing country such as Antigua. The project also aids in the fulfillment of ethical responsibilities of cultural heritage stewards of both Betty's Hope and Antigua. In addition, the vital importance of the interpretation of those resources for the benefit of local Antiguans, neighboring Caribbean Islanders, and international travelers to this cultural heritage site is addressed.

The Visitors Center

The exhibition at Betty's Hope is located within the site's visitors center, a building that has had many purposes over the years, including as a stable and, later, as a cotton storehouse. It is composed of roughly cut stone blocks joined by mortar. There are two doorways and three windows that when open provide a good amount of natural light—an important consideration for exhibit design, especially since there is rarely a generator on site to provide electricity to the overhead lights. One side of the building is closed off to visitors and is used as a storage space. Inside the visitors center, there is a step near the middle that creates two levels, with the floor of the lower level slanting slightly down toward the back door of the building.

The visitors center hosts a large number of visitors to Betty's Hope every year and is a major stop on the tourist route. In addition, numerous local children visit the site during school field trips. Despite all this activity, and even though the label content was rich and factually supported, the previous exhibit was outdated and in a state of disrepair. So in order to uphold the ethical considerations of cultural heritage management, I was invited to redesign the exhibition space.

The Design Process and Materials

This project commenced in the fall of 2013 with the research phase, including content development and the theoretical framework. For my research, I accessed conserved objects from previous excavations at Betty's Hope that were being prepared for exhibition in the Heritage Resources Conservation Laboratory at California State University, Chico.

In the following spring of 2014, while still in California, I wrote all label text for the panels and cases. Exhibit panel content consisted of a history of Betty's Hope Plantation, including the Codrington connection; a brief history of the Atlantic slave trade and slavery at Betty's Hope; the decline of a once-booming agro-industrial enterprise; the process of sugar and rum production at Betty's Hope; an environmental perspective of the site; and the current status of Betty's Hope Plantation, which includes topics concerning the Betty's Hope Archaeological Project as well as other preservation and restoration efforts on site. Since I was unable to visit the installation site, I used schematic drawings to design the gallery layout using Google SketchUp (figure 15.1). This was especially helpful in developing the storyline, determining the layout of the panels and cases, and planning the overall visitor traffic flow.

The panels were designed using Microsoft Publisher, and content sections were differentiated by design and with color in order to aid visitor orientation to the storyline. For example, the background for the history components was rolled parchment, while images of the various products of sugar and rum production represented those respective processes.

Panels were printed on DuPont Tyvek banners and rolled for ease of transport, as all exhibit components and materials were transported via checked airline luggage. Tyvek is also weather resistant, a major factor in the humid tropical climate of the Caribbean, and fairly inexpensive. The banners had pole pockets on the top and bottom with wooden dowels within for hanging the banners and for securing them to the wooden panels. The dowels were affixed to the wooden panels with eyebolts. The wooden panels were painted using paints with low levels of volatile organic compounds to minimize off-gassing, and in various shades of yellow to reflect the light within the structure and to represent the sun on Antigua's national flag.

Exhibit case labels were printed in a clear, readable font, heat-mounted on mat board, then covered in a matte-finished laminate to prevent glare

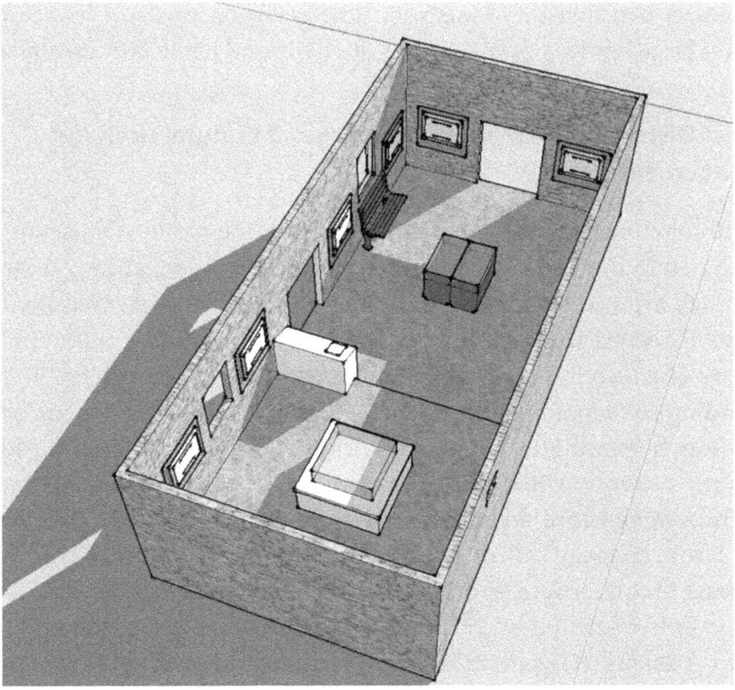

Figure 15.1. Google SketchUp of Visitors Center floor plan.

while protecting them from the humidity. Risers were constructed for the exhibit case interior and covered in Marvel Seal and Oddy-tested linen, which also covered the exhibit case deck. It was vital to produce and fabricate as many of the exhibit components as possible prior to installation, as once on site, there are few opportunities to pick up supplies. The nearest stores are in St. John's, a half hour's drive on two-way roads. Though I had no vehicle of my own, I was able to ride along with Betty's Hope Project participants to St. John's more than once.

In the summer of 2014, I traveled to Antigua and installed the exhibit in the Betty's Hope Visitors Center, as well as a small companion exhibit in the Museum of Antigua and Barbuda, which was a relatively quick and straightforward installation, taking only a single day. At the visitors center, I was able to do most of the installation myself, though I did receive help painting and hanging banners from a local museum employee, and a few Betty's Hope Archaeological Project participants also helped paint. Painting the wooden panels, the exhibit cases, and the model stand was problematic in that there was very little water on site, so I had to bring several large bottles of water to the site daily and paint very carefully.

A Brief History of the Modern Museum and Cultural Heritage Management

Cultural heritage management and the rise of the modern museum grew in tandem in the West and are necessary for an understanding of how to approach a project such as redesigning the visitors center exhibition at Betty's Hope. At the beginning of the twentieth century, museums were centers of education and public enlightenment (Alexander 1997:11), but according to Stephen Conn (1998:31), the epistemological focus on objects was beginning to fade, and by the 1980s, objects could no longer hold the meaning with which they had once been invested. Knowledge was becoming more theoretical and experiential. In 1942, Theodore L. Low wrote an essay called "The Museum as a Social Instrument," which suggests that museums should serve an intellectual middle class. However, events of the 1960s, 1970s, and 1980s changed the meaning of "social instrument" (Low 1942:70).These included the civil rights movement; the immigration of dispossessed African Americans, Latinos, and other ethnic agricultural workers to inner cities; the increased development of white suburbs; and the rise of student and countercultural movements against the Vietnam War (Alexander 2008:221–222). These developments initiated a significant paradigm shift in museology.

According to Kuhn (1962:10), a paradigm shift occurs when there is a change in the basic assumptions or views about a discipline. Applying this idea to museums, a paradigm shift occurred from a focus on collections and elitism to public education and outreach and social interaction. Even though public education has been the mission and intent of American museums since the early twentieth century, a resurgence of interest in visitor engagement prompted a shift to involving museum communities in exhibit planning, programming, and other museum-related activities, while objects continue to provide a substantive pull for visitors.

In his book *Making Museums Matter* (2002), Stephen Weil surmises that to maintain relevancy, museums must have a strong focus on their publics and on reaching their constituencies through civic engagement, as a form of public service. This shift in thought, called the new museum paradigm, is attributed to Weil, who drew upon the original writings of

John Cotton Dana, a museum pioneer who, in his books *The Gloom of the Museum* (1917a) and *The New Museum* (1917b), urged that these institutions be made more inclusive of and relevant to the daily lives of the citizens. Following Dana's reasoning, in "Rethinking the Museum" (1990) and *Making Museums Matter* (2002) Weil advocated for more accountability by museums and public accessibility through a variety of means, such as providing more engaging and educational exhibitions while promoting a sense of empowerment and social identity to the visitor. Now referred to as the new museum theory or critical museum theory, this theoretical approach has emphasized that to be relevant, museums must be more accessible and accountable, reaching out to diverse communities.

In addition, beyond standard curatorship, during the second half of the twentieth century, museology became its own profession with its own professional standards and practices, advocating for preventative conservation, good stewardship of collections, and sensitivity to indigenous communities. The responsibilities in the care of cultural heritage in museum collections have also required a whole new level of knowledge, skills, and training (Pittman 1999:29). This is in addition to the American Alliance of Museums' accreditation process, which includes a formal review process requiring specific standards of performance (Pittman 1999:9).

Meanwhile, this movement was paralleled with developments in the cultural heritage management community, beginning with the passing of the Venice Charter in 1964 and the founding of the International Council on Monuments and Sites (ICOMOS) in 1965 (International Council on Monuments and Sites 2015a). The Venice Charter is a code of professional standards that provides an international framework for the preservation and restoration of ancient and historic buildings. ICOMOS is a professional association that resulted from the Venice Charter, and it works for the conservation and protection of cultural heritage sites around the world and offers advice to the United Nations Educational, Scientific, and Cultural Organization on world heritage sites (International Council on Monuments and Sites 2015b).

In the interests of protecting and caring for cultural heritage and resources, these two movements are inextricably entwined in the twenty-first century. This approach has influenced standards and practices to the storage, curation, conservation, exhibition, and interpretation of objects and sites, as well as to approaches in visitor education and experiences. The net result of this approach is the expectation that museums are held accountable and collections are accessible, while museum staff reaches out to the public by providing engaging and educational exhibits that utilize multiple learning styles. At the same time, cultural heritage site managers do the same with exhibits and interpretive panels among the grounds, which provide meaning and context to what visitors are seeing and experiencing.

At Betty's Hope, and in the Betty's Hope exhibit in the Museum of Antigua and Barbuda, these standards are met by the meticulous excavations of the Betty's Hope Archaeological Project and the exceptional care taken of all artifacts—from excavation to storage and transport—and the conservation treatments performed with permission in the Heritage Resources Conservation Laboratory at California State University, Chico, and a field facility at Nelson's Dockyard utilized in 2017. In addition, great care was taken in the exhibition of artifacts at the Betty's Hope Visitors Center to ensure that they would be preserved for years to come in terms of exhibit design materials, layout, and handling.

Cultural heritage responsibilities are also met through the interpretive exhibit panels, which include text, images, and diagrams. These multiple layers of information were created to be accessible to all types of visitors. The panels are aesthetically pleasing and eye-catching, and they can appeal to a variety of individual learning styles. Text is displayed in a large, clear font style, and labels are written with interpretive flair to enhance interest and learning. Diagrams and photographs are used to create multiple layers of information that can each be accessed individually by the visitor or used in conjunction with one another to add depth to the story. All this was done in collaboration with local Antiguans, as well as with scholars well versed in the history of the site, in order to create an interpretive exhibit that would educate and entertain both local and international visitors.

Theory in Practice: A Ticket to Interpretation

As new museum theory advocates for more accountability by museums and more accessibility, the exhibition at the Betty's Hope Visitors Center and the Betty's Hope exhibit at the Museum of Antigua and Barbuda both address accessibility through exhibit text, panels, layout, and archaeologically recovered objects. At the Betty's Hope Visitors Center, this includes arranging the panels so the storyline reads from left to right, but also

Figure 15.2. New exhibit panels installed, 2014.

allowing for visitors to view any panel out of order to absorb a smaller, more digestible part of the story (figure 15.2).

An appropriate font size and style was carefully chosen to allow for ease of reading. Most importantly, it was necessary to take into consideration visitors' preconceived notions prior to entering the exhibit. Considering that visitors range from local Antiguan schoolchildren to traveling tourists and that they come to the exhibit with varying levels of education and from diverse socioeconomic backgrounds, writing accessible label text was imperative. This includes keeping labels short when possible and in plain language, generally between a sixth- and eighth-grade reading level (Serrell 1996:97), and employing a constructivist learning perspective, which allows visitors to construct knowledge and meaning from their own experiences instead of assuming they were blank slates or empty vessels to be filled with knowledge (Hein 1998:22). Although multimedia and interactive exhibits would have been engaging additions, they would not have fared well in the humid, partially exposed structure. In addition, there is not a technician on site during open hours to repair these types of exhibits should they malfunction or break down. All this was done with local Antiguans in mind.

Accountability was addressed through the rigorous scholarship behind the development of exhibit content. This included consulting scholarly articles and texts and working personally with Georgia Fox and Reg Murphy, scholars of Antiguan history, archaeology, and culture. Reg Murphy was an ideal candidate to peer review all label text for content, depth, and cultural sensitivity as someone of Antiguan descent who currently lives on island. These perspectives combine to tell a complete and well-rounded story through interpretive text. Working with both local Antiguans and content scholars was of the utmost importance, not only because this fills the ethical requirement of museums and cultural heritage managers to collaborate with descendent communities, but also because it is imperative to portray cultural sensitivity and credibility by engaging with people who live Antigua's history every day and scholars who are engaged in these topics.

Cultural Heritage Management in Practice: The Final Road to Interpretation

The renewal and installation of the exhibit at Betty's Hope Visitors Center is an important component of the cultural heritage management at this site. Great care was taken in the exhibition of these and other artifacts at the Betty's Hope Visitors Center and the Museum of Antigua and Barbuda to ensure that they will be preserved for years to come. I have gone further to protect these assets by employing methods to minimize off-gassing within the exhibit cases, which could pose a threat to the perpetuity of these objects. This is yet another example of accountability.

Reg Murphy and other local museum employees and volunteers take great care to maintain the property as time and funds allow. At the moment, there is no one scheduled full time to oversee Betty's Hope all day, every day; however, there is a local Antiguan who unlocks the gates and doors to open up the site and the visitors center, keeping it and the on-site restroom facilities clean and maintained. He also locks up again at the end of the day and delivers entrance fees (donations) to the Museum of Antigua and Barbuda.

Interpretation: The Destination

Interpretation is an important component of the cultural heritage management activities at Betty's Hope. In addition to the interpretive aspects previously discussed, other interpretive components add to the visitor experience, including signage throughout the property, as well as a constant influx of well-versed and well-trained tour guides employed by private and for-profit companies, each of whom offers a unique and personal perspective of Betty's Hope. Off-site interpretation includes websites run by the Museum of Antigua and Barbuda and travel agencies, as well as the small exhibit within the Museum of Antigua and Barbuda.

In trying to assess the level of interest and helpful feedback regarding the completed exhibition, I had the opportunity to have informal conversations with the tour guides who came out to the site on a frequent basis, especially the guides from a local tour company, Island Safari, who come out to the site almost daily in open-air jeeps as part of their island excursions for tourists.

On the whole, the exhibit was well received by these tour guides, local grade-school teachers, and others. Additional stakeholders included museum personnel from the Museum of Antigua and Barbuda, local news crews for television and print journalism, government ministers, and Reg Murphy and Georgia Fox. During the summer field seasons of 2016 and 2017 and a visit in 2019, Fox reported that the exhibition at Betty's Hope Visitors Center was found to be in pristine condition, verifying that the materials chosen are proving to be durable after thousands of visitors and the vagaries of local climate and weather.

During installation, many visitors came through the visitors center. Returning international tourists were pleased and impressed with the upgraded exhibit, offering compliments and congratulations. At one point, a visitor from San José, California, was looking at the artifacts within the exhibit cases, which had yet to have their interpretive labels inserted. Initially, she told me she found the conserved metal objects quite boring, but when I explained the need for objects to have interpretive text to reveal the stories associated with them, she became much more interested. She seemed to enjoy our conversation as I explained the meaning and history behind the objects and how they can contribute to a better understanding of the archaeology of the site, which can then paint a picture of what life was like for the inhabitants of Betty's Hope and other plantations like it.

Later that same day, I was able to have a conversation with two local Antiguan tour guides, who regularly bring guests to the site, and two contributors to the Betty's Hope Archaeological Project. During this powerful conversation, we discussed the history and importance of Betty's Hope and cultural heritage, as the two local tour guides explained Antiguan history with great pride and knowledge. All agreed about the importance of sharing research at Betty's Hope and places like it with informed tour guides and the public. Finally, travel reviews posted on websites such as Trip Advisor seem to be favorable. The hope and intention is that these positive reviews will garner more interest in, and a positive association with, the site.

Conclusion

Despite the perception of the success of the exhibit renewal, Betty's Hope could benefit from greater attention to upkeep of the site. Bees, weathered signage, invasive plant species, and free-roaming livestock pose ongoing challenges to site integrity and maintenance, all of which need to be addressed. The renewal and installation of the exhibit at Betty's Hope Visitors Center is one important step in the greater heritage management of this important site and other historic and prehistoric sites on Antigua. By showing the public the value we place on the stewardship of sites like Betty's Hope, we hope that visitors will leave Betty's Hope with a greater understanding of our shared histories and of the critical role island nations like Antigua played on the world stage in the making of early modern history.

16

Using UAVs to Manage Archaeological Heritage

A Multiscale Analysis Approach

ERIN FRIEDMAN, CORY LOOK, AND MATTHEW BROWN

This chapter explores the use of UAVs (unmanned aerial vehicles) for the use of archaeological investigations and heritage management at the historic sugar plantation of Betty's Hope, Antigua. While the acquisition of low-flying aerial imagery, such as kite photography, has been common practice within archaeological research, recent software innovations coupling photogrammetry and UAV technologies are providing new tools for exploration. Two different approaches for UAV acquisition are explored in this chapter: the first for use within archaeological excavations and the second for studying the landscape. Both have particular implications for heritage management, as structure-from-motion (SfM) methodology coupled with aerial imagery can be used to produce an accurate three-dimensional (3D) surface model of the site that is akin to site scanners and light detection and ranging (LiDAR) technology. The important differences between and limitations of these technologies will also be discussed.

Our work adds to the growing body of literature on UAV and photogrammetry applications within archaeological research and heritage management. Climate-change impacts on heritage sites have become a central concern in global and regional climate reports that inform policymakers and heritage managers, and as such, this case study has potential implications for managing and studying ecological and climatic risks that threaten heritage sites.

Documentation of archaeological sites is integral to the management of cultural heritage. Methods of recording sites have transformed over time with the introduction of different technologies to document features. Remote sensing, geographic information systems (GIS), and photogrammetry techniques to archive archaeological information complement current archaeological methods of recording features and site landscape. Simultaneously, the integration of these techniques has also commanded an evaluation of the potential opportunities they offer for short-term and long-term management of heritage. In the study, we evaluated the barriers UAV technology overcomes and the opportunities it offers for recording of archaeological sites and heritage management. Recording methods utilizing UAV, GIS, and photogrammetry were implemented to survey Betty's Hope at site scale and landscape scale. Concerning cultural heritage management of Betty's Hope, this site demonstrates the need for tools that provide rapid availability of information for cultural heritage manager decision making. The threats Betty's Hope faces require sustained assessment and evaluation of the site.

Sustained site assessment and evaluation requires personnel and technological support, which can be expensive and time consuming, stressing the already limited capacity of the Antigua and Barbuda national parks authority. We have combined the accessibility, precision, and reliability of UAV technology with traditional landscape archaeological methods to aid in rapid and continual assessment. This chapter will show how a sustained assessment combining these methods can benefit heritage managers to record sites on multiple scales and to monitor anthropogenic hazards (e.g., land clearing).

Background of Betty's Hope Observed Impacts

The presence of some extant structural remains at Betty's Hope has made the site vulnerable to destructive forces such as weather and climate, invasive plant species, grazing domesticates such as goats and sheep, and human activity, which all threaten the site. The two windmills, the still house, the remains of the manager's house, the four large cisterns, and the scattered ruins are all susceptible to damage over time. These conditions are further compounded by recent theft from the site of limestone and historical farming equipment, which is sold for metal export. Clearing land for safety reasons has dominated land clearing practices in Antigua

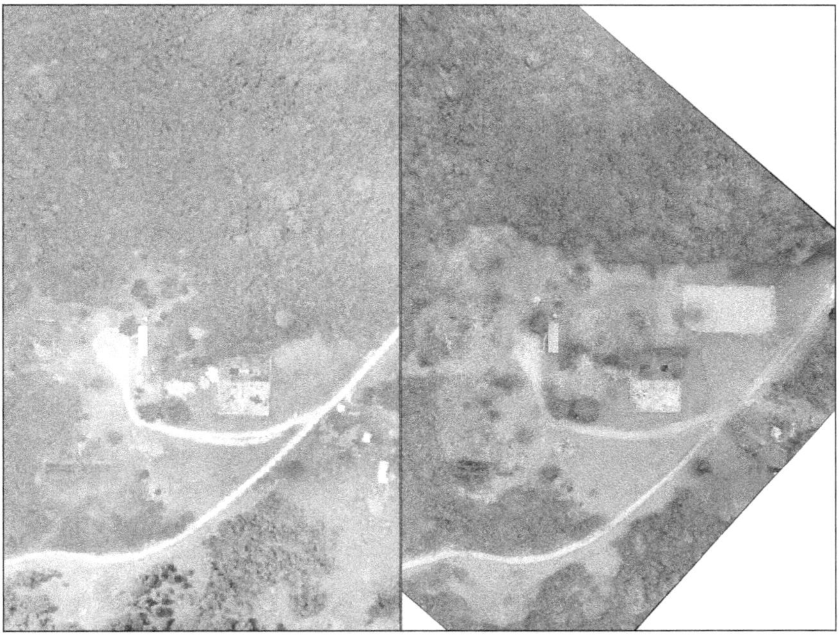

Figure 16.1. Betty's Hope site before (*left*) and after (*right*) backhoe operations.

and was one of the reasons for a substantial clearing of the site in 2014, when a pair of visiting tourists generated a dispute concerning their experience at Betty's Hope. The response by the government was to address their concerns with bulldozing the northeastern section. The bulldozing occurred swiftly after the incident, without consultation from local archaeologists (see figure 16.1).

The destruction, while dramatic in appearance, removed the top 20 cm, exposing habitation deposits, which provided an opportunity for archaeologists to document and investigate this section of the site. A quick surface collection and mapping of artifact concentrations helped redirect that season's excavations to this area in order to salvage materials now threatened by erosion. The clearing of sites in Antigua has been followed by the rapid regrowth of vegetation, limiting the amount of time for documentation within the scope of that season. As such, aerial documentation of Betty's Hope became an immediate priority for both archaeologists and heritage managers.

Archaeology, Aerial Photography, and UAVs

Aerial photography in archaeology was first used in 1880 and gained popularity in its implementation following O. G. S. Crawford's work with the Royal Air Force to photograph British sites such as Stonehenge in 1919 (Reeves 1936). The use of aerial survey for high altitude, and helikites, balloons, and poles for low-altitude imagery of archaeological sites, provides two-dimensional (2D) images for documentation at different resolutions. Aerial survey by plane is expensive and time consuming in data collection and processing and may not be appropriate for certain types of data collection. Low-altitude aerial photography (LAAP) is better suited for site-scale documentation as there are many options available to capture images at specific altitudes for relatively low costs. Although technology such as kites, balloons, and helikites has proven to be quite useful to archaeologists over the years for LAAP, weather conditions and limited mobility restrict the positioning capability for desired site images (Verhoeven et al. 2009) .

Concerning archaeology, UAVs are able to provide flexibility and greater control by the user to capture complicated structures of sites. A variety of UAVs of different sizes and cost and ranging from rotary to fixed wing are available on the commercial market. The piloting of a UAV can be done with user control by remote or a pre-flight plan that is programmed through applications on mobile devices or laptops. The most popular UAVs among researchers are quadcopters (four rotors).

There are models with their own cameras and those whose cameras are purchased separately, dependent on resolution needs. Quadcopters have a small payload (the maximum weight they can carry) of about 1,300 g. Larger UAVs, such as octocopters (eight rotors), have heavier payloads. The payload of a UAV is extremely important, as it can determine what can be attached, which varies depending on what kind of data is needed. Heavier add-ons include higher resolution cameras such as single lens reflex (SLR), thermal cameras, and LiDAR rigs. In particular, as the weight of miniature LiDAR rigs has decreased, their experimentation with UAVs has increased.

For our investigations, over the course of two weeks, we employed a quadcopter UAV at Betty's Hope to capture 2D and 3D imagery of the site using photogrammetry techniques to provide a detailed, high-resolution digital record of site structures. The UAV's portability and capability to be stationary within high-speed winds at Betty's Hope was useful for LAAP of the site and provided rapid recording for heritage managers. The relative stability of the quadcopter stems from the on-board GPS and accelerometer, which work together to help minimize drifting. The camera is mounted to a three-axis gimble, which is also used in professional film making, and provides additional stability while in flight. Digital records from orthophotos and 3D models of Betty's Hope were meant to generate a digital record for future cultural heritage management, a growing trend within archaeology (De Reu et al. 2013). Orthophotos are aerial photographs that have been geometrically corrected to fix warping and other types of distortion that occur during aerial photography. This allows the orthophoto to be used as a base map for true scale measurements. Helikites use SLR cameras with minimal optical distortion; however, the lack of position control and integrated software render this process problematic in producing orthorectified images for base map measurements.

Methods

This research focused on the use of low-cost and accessible UAV equipment. We decided to use a commercially available DJI Phantom 3 UAV along with the GoPro Hero 3 Plus. The issue with low-end commercial UAVs is that they have a limited payload that prohibits SLR cameras. Our reasoning was to test a popular and available UAV with a low learning curve for aerial image acquisition. Confronting these shortcomings requires preprocessing of imagery in order to compensate for the fisheye effect of the lens.

We tracked the altitude, pitch, speed, and GPS location of the UAV and conducted automatic flight patterns. The process for attaining orthophotos and 3D imagery of the site involved fieldwork that required ground control points (GCPs) and LAAP. Software for flight planning was integrated along with site surveys in order to minimize encounters with obstructions or private property. A rangefinder was used to calculate heights of vegetation, standing structures, and electrical poles to establish the lowest safe altitude. Post-fieldwork involved rectifying the optical distortion from the photos and composing photos for creation of a 3D model.

Data Collection

Fieldwork involved recording GCPs and taking images of the site. The team created a plan that incorporated separate flight plans, analysis, and data collection for both scales of the site. The excavated great house and kitchen walls and floors, and the area surrounding it (site scale), were our objectives for site capture. While the great house excavation area would be included in the site-scale imagery, there was concern about the resolution of the great house area itself at this scale, as details of the exposed excavated walls and floors were important to capture.

The GCPs chosen were structural features that could be visible and identifiable from aerial imagery. GCPs are significant to record as they are used to georeference the orthophotos (i.e., referencing imagery to points recorded in physical space within a specific geographic coordinate system for true measurement). Each GCP was marked with a round orange disk. Unique target patterns were printed on each disk, with a small hole drilled in the middle to place the stadia rod for use with the total station into the middle of the disk. For both the great house imagery and site-scale imagery, the round orange disks were distributed across those sites. A total station survey and stadia rod with prism attached were used to record the digital elevation points of the GCPs with ±3 mm resolution. Fifteen points were recorded for the great house and 24 points recorded at site scale.

The team deployed a UAV manually during the early morning to have the least amount of shadowing in the images. Shadows can cause the loss of spectral information in images, affecting parameters of pixels that are used for digital terrain modeling (Dare 2005) and obstructing important features for analysis. The UAV flight plan was managed in a way to obtain overlapping images. The overlapping of multiple images is important for deriving 3D imagery from the site as the redundancy allows multiple features to be extracted (Westoby et al. 2012). Images were recorded at one photo per second.

Concerning the site-scale flight plan, the UAV was flown at 180 m above ground level for one round and 210 m above ground level for another round. The purpose of this was to experiment with pixel resolution and see which height offered the best resolution for site-scale recording. For the great house, the UAV was flown at 50 m above ground level. For each set of imagery, two flyovers were completed for each.

Post-Processing

The post-processing of UAV images from Betty's Hope began with sorting photos that could be used for orthophoto and 3D model creation. This was necessary as the quality of the orthophoto and 3D model is dependent on the quality of original imagery (De Reu et al. 2013). During the photo screening process, images that were blurry, did not contain appropriate angles of the site (i.e., images captured before UAV takeoff), or contained shadows were eliminated. Afterward, the screened photos were treated for optical distortion through batch processing, allowing for single image rectification to be completed in an automated fashion across hundreds of images, saving a great deal of processing time for developing two different models.

For the 3D Model

Following the screening process, the photos were aligned to create a sparse 3D point cloud from the orientation and position of the photos. A sparse 3D point cloud is composed of a lesser density of points relative to the digital surface model as a result of this process using lower resolution imagery (Harwin and Lucieer 2012). Therefore, 59 out of 61 photos chosen for the great house were aligned successfully and used for the 3D model. This resulted in the successful aligning of 265 out of 295 photos of Betty's Hope and was used to create the 3D model. Prior to creation of a 3D model, the alignment of photos must take place. A digital surface model has to be created prior to creation of the orthophoto (figure 16.2).

Timeframe

The team spent approximately 10 days in the field and five days in post-processing to produce the final products. The small production time was due to the post–field processing using the SfM approach. SfM is an algorithm from the computer vision field that creates 3D representations of imagery based on multiple overlapping images of the same object. The camera does not have to be calibrated; instead, the position of the imagery based on the overlap is used by the SfM algorithm to develop a 3D point cloud. This differs from traditional photogrammetry that relies on specified targets and camera calibration before data collection.

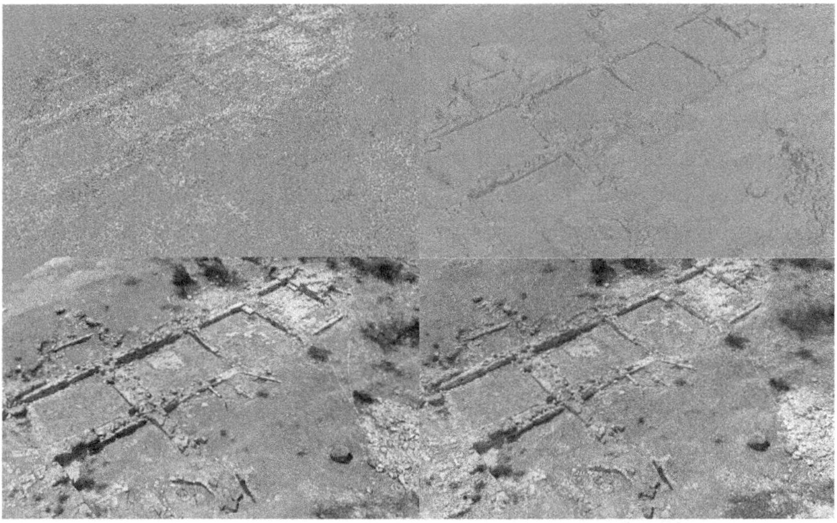

Figure 16.2. Phases of 3D modeling for the Great House. Sparse 3D point cloud (*top left*); shaded mesh (*top right*); smoothed texture (*bottom left*); rough texture (*bottom right*).

Processing of the orthophotos and 3D imagery used Agisoft's PhotoScan, photogrammetry software popular for its use among archaeologists for capturing 3D imagery of artifacts. This software uses SfM methods to automatically correlate matching features from overlapping images by estimating relative camera locations (Snavely et al. 2008).

UAV Applications for Archaeological Excavations

Excavations of the great house by Georgia Fox and California State University, Chico, were documented and recorded using total station measurements and overhead imagery taken with an SLR digital camera. We avoided using a fisheye lens because of its warping of imagery along the edges, resulting in poor accuracy. Each photo was georeferenced using Esri's ArcGIS software using GCPs that were visible in each photo and referenced accordingly. This process allowed for accurate and speedy documentation of architectural features and structural remains (figure 16.3).

The use of digital recording should not entirely replace traditional hand-drawn context or recording sheets, but rather should complement an archaeologist's toolkit. Many archaeological investigations rely solely on hand-drawn plans that are then scanned and entered into a software

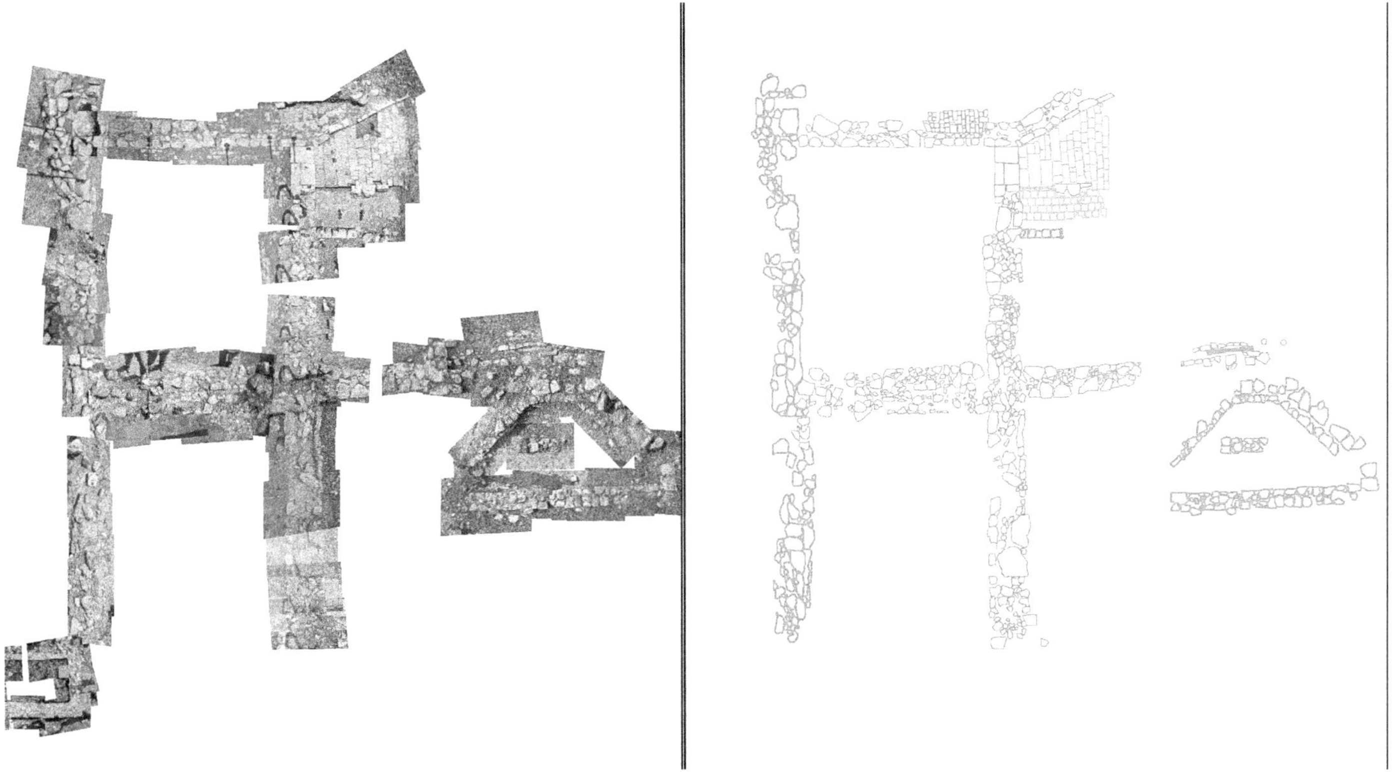

Figure 16.3. Georeferenced overhead SLR images (*left*); digitized architectural plans based on overhead images (*right*).

program like ArcGIS or AutoCAD to be digitized. Prior to the integration of this approach, the bulk of processing and synthesizing of spatial data and overall site plans was conducted after the season, providing feedback on planning for subsequent seasons. The advantage of using georeferenced overhead imagery is tied largely to time management, where archaeological planning shifts from in-field to nightly post-processing work.

Additionally, this method saves a great deal of valuable field time and places a larger emphasis on excavation and in-field planning that would otherwise not occur until periods of post-excavation. Site plans were updated each evening, with field notes transcribed into coded attributes for each feature. Stone structures could therefore include their relative size, parent material, and presence of plaster or mortar, while spatial relationships between a variety of features and assemblages could be analyzed to generate datasets used to make decisions that may have previously been implemented in subsequent seasons.

The integration of a UAV to collect overhead imagery did not entirely replace this process; rather, it increased the scope and scale of site documentation and further integration of landscape data. Excavations at Betty's Hope encompassed an area of 250 × 170 m, while landscape surveys covered an area approximately 360 × 380 m. The topography of the site along with thick vegetation meant that obtaining lines of site with the total station were largely impractical. The same dense vegetation surrounding the site prohibited the use of a differential GPS by blocking satellite signals for triangulation.

Prior to the use of LAAP, we relied largely on handheld GPS devices with an accuracy between 5 and 25 m to record excavation units, shovel test pits, and surface scatter throughout the entirety of the site. The great house excavations, measuring 32 × 21 m, were recorded using overhead imagery collected with a boom-mounted digital SLR camera and total station. A total of 87 SLR overhead images were taken, each with its own set of GCPs necessary for georeferencing. Revisiting the great house excavations using UAV technology demonstrated the potential for future archaeological investigations. A total of 61 photos were collected at different camera heights, positions, and angles over the course of 45 minutes of flight time.

These images were collected at heights between 7 and 15 m and were processed using SfM, resulting in a single orthorectified aerial image

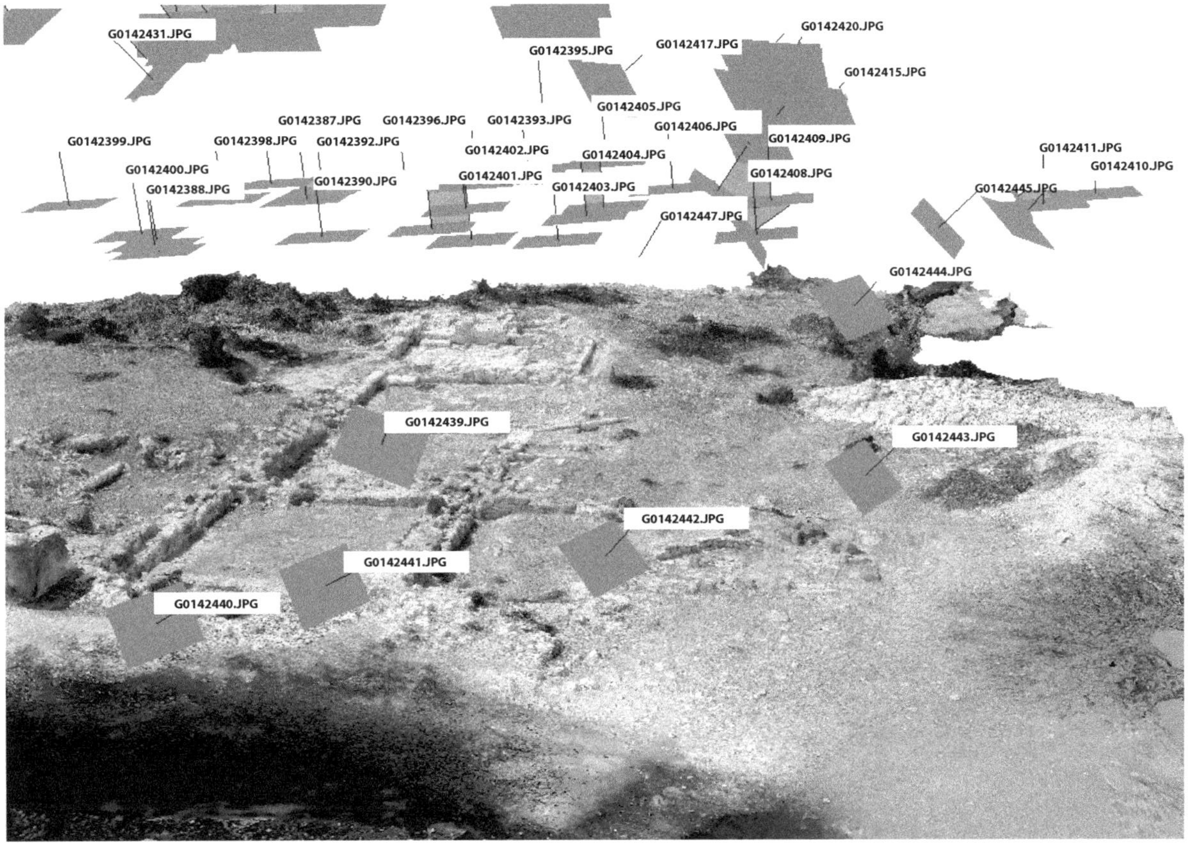

Figure 16.4. Camera angle positions from UAV flight over the Great House.

covering the entire excavation area of the great house. A single series of GCPs were collected with a total station and were used to georeference the imagery. The orthorectified image constructed from the UAV photos resulted in a pixel resolution of .37 cm. A number of studies have shown that using these methods produces datasets comparable to LiDAR (see figure 16.4; Fonstad et al. 2013; Mancini et al. 2013).

The resulting imagery provided an expanded overview of not just the excavation area, but also unexcavated areas adjacent to these spaces. These low-flying aerials showed clear changes in vegetation and geometric alignments associated with buried structural features that were not previously visible or clearly delineated from the ground perspective. This technique is commonly used with aerials acquired from costly commercial flyovers of island-wide collections. Aside from cost, the quality of aerials is not consistent throughout the islands, stemming largely from cloud cover and shadowing. The resolution of high-flying aerial images collected from Antigua in 2010 had a pixel resolution of 1 m, in large contrast to the .37 cm pixel resolution collected with low-flying aerial acquisition. The 1 m aerial was also inadequate for the aerial analysis of structural features for the great house (figure 16.5).

The method of image acquisition necessary for orthorectification is consistent with imagery necessary for constructing accurate 3D models of excavations and landscapes. Using photogrammetry software, the 61 photos were acquired using a UAV-generated point cloud made up of 129,832 points. The Agisoft software allowed the user to define the accuracy for each model. Our research found that structural features worked better under high resolution, resulting in increased detail and accuracy for structural features. When dealing with heavily vegetated landscapes, lower resolution models resulted in more accurate models. This may have been a byproduct of how the software aligns imagery to construct point clouds, with moving vegetation posing the largest problems in alignment. Photogrammetry fares better when photos can be properly aligned with distinct and stationary features or targets. This may be overcome with the integration of larger targets, perhaps by laying out a series of large non-reflective fabric in or adjacent to open areas of heavily vegetated landscapes to create these stationary features.

The use of 3D models at the excavation scale meant that units could be revisited without being on site. During the documentation and interpretation of units, this provides archaeologists with new tools to study

Figure 16.5. Resolution of 1-meter aerial (*left*) is lesser than that of the UAV acquired aerial (*right*).

stratigraphic relationships, particularly after they have been destroyed through the process of excavation. Three-dimensional models of excavation areas provide additional details that are sometimes missed in the photo record or when particular angles of features are needed but images are not taken. This tool is remarkably useful during the post-excavation analysis and planning phase for subsequent seasons. These models have been used to capture excavated structures, burials, and anthrosols. With respect to museology studies, these models provide a valuable record for both conservators and interpretation specialists. With the growth of 3D printers, excavations could be replicated at a variety of scales for museum exhibits.

Applications for Landscape Survey and Remote Sensing

As previously mentioned, the use of low-flying aerial imagery combined with total station measurements helped generate orthorectified imagery to accurately map excavations, shovel test pits, and surface scatter across the entirety of the site. Prior to the integration of UAV technologies, aerial imagery was acquired largely from island-wide surveys conducted by Antigua for development and planning strategies. While sets of aerial images were taken since the 1940s, many of the flyovers were limited to specific regions of the island, with many images obscured by cloud cover. Nonetheless, these images proved incredibly valuable during the early stages of planning and survey work at Betty's Hope, particularly the georeferencing of estate maps and historical surveys for the island. These aerials captured the different phases of land cover and land-use change, resulting in the identification of old boundaries and roads present in old estate maps. While standing structural features were visible, the resolution provided only an approximation of its location and could not be used to accurately record archaeological features and site plans. Findings from our case study site of Betty's Hope suggest that UAV technologies hold particular applications at the landscape and broader site scale.

The resolution obtained using low-flying aerial imagery was adequate for the identification of structures and mapping of exposed features. Once the necessary photos were processed to produce an orthorectified aerial, the combination of GPS, total station, and landscape surveys helped overcome many of the challenges related to vegetation cover and line of sight that plague surveying instruments. Structural features, once ground truthed, could be identified from high-resolution aerial imagery by identifying vegetation changes associated with structural outlines. Quick measurements of structures and features could be made using GIS software, once the orthorectified image was properly georeferenced.

The increase in resolution meant that additional details not previously visible were identifiable, particularly abandoned roads, walking paths, and collect ponds covered by dense vegetation. A total of 295 low-flying aerial images were used to produce an adequate orthorectified aerial of the site. A total of 481,413 points were generated to produce a 3D model of the landscape with 12 GCPs clustered around the sugar works area. Once the model was properly oriented and the orthorectified image georeferenced, the final landscape aerial had a resolution of 2.91 cm/pixel. Images were acquired between 50 and 150 m, primarily to avoid large trees and powerlines scattered throughout the site.

The 3D modeling of Betty's Hope sugar plantation holds particular potential for creating a more accurate assessment of structural features, particularly in documenting the gradual melting of archaeological sites due to weathering, erosion, and time. Recently exposed areas, such as the great house, have begun to break down and degrade. Grazers, tourists, and local patrons have contributed to the inevitable decay of the site as well. These models can provide teams with limited resources, an alternative method in capturing and documenting architectural details. While vegetation and land cover maps can be recorded in greater detail, the integration of 3D modeling holds potential in generating other metrics, such as vegetation growth and vegetation volume measurements.

These models could be used to track the impacts of land clearing and vegetation change surrounding cultural heritage sites. Densely forested areas that have been cleared do not return to their original form and are often replaced by dry scrubland vegetation. In periods of drought, these newly cleared lands are additionally at risk for erosion, which these models may be useful in tracking. In some instances, 3D models can be used to monitor the effects of erosion mitigation as well as surrounding cultural heritage sites.

The recent bulldozing of the northwestern section of Betty's Hope has resulted in significant vegetation changes in the last three years. What was largely a mix of logwood and lignum vitae forest has been replaced by low-lying scrub. The introduction and invasiveness of vegetation growing out of cleared lands have been documented by the Environment Division in the National Parks at English Harbour as an ongoing problem throughout Antigua. Alternatively, there is also a degree of uncertainty related to the use of plants for mitigating erosion.

Attempts to mitigate erosion along the water catchment at Wallings, located in the southwest of the island, illustrate both of these problems. Fever grass was planted in 1902 to help mitigate the effects of erosion around Body Ponds, with a particular emphasis around Wallings. A 1966 soil survey of Antigua showed that fever grass was being effectively managed due to the continuing sugarcane monoculture still taking place throughout the island. However, the abandonment of farms allowed fever grass to spread more readily, and contradictory to early land-use managers, fever grass seems to exacerbate soil erosion as its roots lift the surrounding soil into raised clumps. This is especially problematic in areas with steep slopes and gradual inclines, which, combined with drought conditions followed by the rainy season, results in the ongoing loss of topsoils. Fever grass is also flammable, and after wildfires, it is particularly opportunistic.

By 1987, fever grass had spread into areas surrounding the sugar plantation of Greencastle Hill, located in the central west portion of the island, and has recently begun impacting the site from the effects of both erosion and wildfires. Subsequent 3D models of this region will be used to help contribute to our understanding of how land-clearing practices affect cultural heritage site degradation. Land clearing is a common practice throughout the world prior to development projects or land sales. While direct destruction of cultural heritage sites has been well documented around the world, vegetation and land cover changes have not been studied in depth.

Responding to Threats to Cultural Heritage

Since the writing of this chapter, we have been involved in a number of separate but related projects ranging from coastal erosion, tourism development, and pre-Columbian site mapping. The approach described in this chapter served as a structural base for documenting, recording, and subsequent modeling of various landscapes and environments. These approaches are particularly applicable to institutions with relatively small staffs and limited resources; however, there is a moderate learning curve necessary to produce consistent and reliable products. These models have been used to demonstrate to stakeholders the extent of loss and destruction of coastal environments and cultural heritage sites. Architectural assessments were also conducted in order to assist the government of Spain in collaboration with Antigua to estimate costs of restoring the military fort at Monk's Hill in Antigua. Documentation for this survey was completed within less than one week.

Climate impacts on cultural heritage sites are an area of increasing global concern. Uncertainties related to the timing and frequency of extreme weather events have made it increasingly challenging to plan for appropriate mitigation efforts. High-resolution digital elevation models are tantamount in assessing threats from water inundation and storm surge. Given that the current UAV technology can be stationary with local environmental conditions (e.g., wind), deploying a UAV to rapidly record data, combined with photogrammetry techniques using SfM methods that do not require camera calibration, provides rapid recording, data processing, and reporting for heritage managers to make decisions about response. In addition, the techniques presented in this chapter provide methods for tighter chronological control, having continuous data collection before and after impact.

While Betty's Hope is not threatened by water inundation, it is vulnerable to extreme hurricane-force winds. During the fall of 2017, Antigua and Barbuda experienced a category 5 hurricane with wind gusts greater than 160 miles per hour. While Antigua did not experience a direct hit from the storm, it did sustain some structural damage. Digital records as described in this chapter would be invaluable in future attempts to restore the plantation for future heritage tourism.

Afterword

Final Days at Betty's Hope

GEORGIA L. FOX

Am I not a man and brother?
John Greenleaf Whittier

The archaeological evidence for the last days at Betty's Hope is ephemeral and best demonstrated through the documentary record. What is known is that distant rumblings of impending abolition sent planters into high alert, and the Codringtons were no exception. The Abolition Slave Trade Act of 1807, which abolished the Atlantic slave trade for the British, sent a clear message, although it would take another 27 years before slavery ended on Antigua. Furthering matters, compulsory registration of slaves on Antigua was codified in 1817, and pressure from the Moravian Church and other religious affiliations called for the education of the enslaved, heightened by the growing number of free blacks by 1820 to almost 4,000 on island (Dyde 2000:114, 117).

Requests at Betty's Hope for freedom were made as early as 1771, when Betty's Hope manager Samuel Redhead asked to purchase the freedom of Sarah Bullock (NAAB Codrington Papers:CC36; see chapter 4 of this volume). Such requests continued into the 1800s, when an overseer at Betty's Hope requested permission in 1816 to purchase the freedom of the children he had with one of the plantation's slaves, a Mustee woman (NAAB Codrington Papers:CC301), but generally, such requests were mostly on a case-by-case basis, which offered little threat to an established way of life.

In contrast, the thought of wholesale abolition sent Betty's Hope's managers into a tizzy for what they perceived to be the unfairness and unreasonableness of it. In a letter from plantation attorney John Osborn to Christopher Bethell Codrington dated October 23, 1823, Osborn bitterly blamed Parliament for hanging the planters out to dry and the Methodists for their "subversion" (NAAB Codrington Papers:C28). Working himself into a lather, Osborn complained that "our lives and properties have been sacrificed. It is lamentable to see our best rights infringed with impunity for parliamentary interests. . . . Surely government cannot accept that we will enlighten or prepare our properties for emancipation, a total loss to ourselves without having a compensation in view."

Talk of a post-emancipation world agitated matters on Antigua. Chief among decisive actions taken was a law passed in early 1824 by the Antigua Assembly prohibiting the Sunday market. This exacerbated unrest among the enslaved population island-wide. Public demonstrations by Afro-Antiguans to protest the law were accompanied by intentionally set cane field fires beginning March 20 of that year, many of them in St. Philip Parish. These actions were swiftly met with militia sent out to round up the perpetrators (Dyde 2000:127). It is unknown if Betty's Hope had cane fires, being in nearby St. Peter's Parish. Still, it was a tense situation. Planters and their managers complained about the "rebellious behavior of negroes." In a letter dated October 6, 1829, Betty's Hope Plantation manager R. Jarritt, Osborn's successor, noted "the revolt of the Negroes at Betty's Hope and the Garden [Estate] was sent home to the colonial office." He further stated that "if the Negroes had not had such instigation . . . they would not have gone to the lengths they did" (NAAB Codrington Papers:C29).

Eventually, as the date of emancipation grew closer, resignation set in with the impending change. Realizing the inevitability of the shifting winds, Christopher Bethell Codrington took out a large advertisement endorsing emancipation, dated August 24, 1832 (NAAB Codrington Papers:E38). Parliament's Abolition of Slavery Act of August 28, 1833, set the date to be officially August 1, 1834, for all of the British colonies (Dyde 2000:131). In September of 1833, planters, particularly those faced with mounting debt, protested the act at Antigua's courthouse (Dyde 2000:131). In the end, however, they lost the fight, and Antigua's 30,000 newly liberated peoples were free to seek out a new life, but for many of them, it was not smooth sailing (Lightfoot 2015:1).

Post-Emancipation at Betty's Hope

Recent scholarship on the slave trade and sugar economy appears to support Eric Williams's thesis that slavery helped Britain's elites amass a vast surplus capital that enabled them to bankroll the Industrial Revolution (see Blackburn 1997; Williams 1994). After enslaved peoples were free, the planters demanded payouts for their "loss." Christopher Bethell Codrington, in fact, received a payout of £4,920.9.0 for Betty's Hope on October 19, 1835, for 299 slaves (University College London [UCL] 2017). Codrington received an additional payout for his other Antigua and Barbuda estates, totaling £20,089.51.4 for 1,922 slaves. Adding Betty's Hope, this came to a payout of approximately £25,000 for 2,221 slaves, a hefty amount for the time, over half a million dollars in today's currency. This does not include payouts for the other Codrington estates on Tobago, Jamaica, and Grenada, for which other family members received compensation (UCL 2017).

Without enslaved labor, planters applied every ounce of their collective efforts to consolidate power and exercise control over the newly freed population (Williams 1994:4). As one informant stated, the *bakkra* (white planters) "use the militia to keep people in check and the militia would have the back up of the magistrates and the jail-house and the government. No way for us to fight back—it was the like worm going against nest of ants—for the bakkra was the militia and the magistrates and the jailhouse and the government" (Smith and Smith 1986:46).

The payout allowed some planters to expand and update their operations. Among these were the Codringtons, who wasted no time on this front as they advanced to steam power (see chapter 7). Yet as Dyde (2000:167–170) observes, the upgrade to more modern machinery was no real panacea for a system that was limping along at this stage. Two factors contributed to less-than-ideal conditions. First, the Sugar Duties Act of 1846 essentially removed the preference for British sugar. The second factor was the ongoing debt accumulated by the planters, making it possible for only the wealthier planters like the Codringtons to afford the cash for specialized equipment.

A powerful earthquake struck on February 8, 1843, terrifying Antigua's inhabitants, with the "laborers cottages for the most part require considerable repairs, and several to be rebuilt . . . [and] the Dwelling House . . . is in thorough repairs" (NAAB Codrington Papers:E23). Not quite four years later, the island was clobbered by a massive hurricane in 1848, leaving the island in disarray.

For the most part, the plantations were dismal places to work, and freed peoples who could not stomach another day on the old plantation pursued their own destinies as fishers, porters, artisans, domestic workers, peddlers, laundresses, and other occupations. By 1844, Antigua's population of over 36,000 inhabitants were engaged in household duties (13,503), served as domestic servants (some 2,017), worked as mechanics and artisans (2,521), or were involved in miscellaneous activities (over 1,200) (NAAB Codrington Papers:E23). After emancipation, missionaries descended upon the region, with Antigua's 5,000 children being educated by the Anglican, Methodist, and Moravian Churches, resulting in a legacy of having one of the highest literacy rates in the Caribbean today (Dyde 2000:151).

As with the other British sugar islands, the preoccupation with a labor shortage prompted the planters to push back by enforcing the Laborers' Contract Act of 1834, which introduced stringent work contracts for free laborers. This strategy was rejected by London, but others took its place. Part of the troubles stemmed from the firmly rejected six-year apprentice system that was adopted throughout the rest of the British Caribbean but not on Antigua. Meanwhile, Afro-Antiguans formed their own villages at the edges of plantations, such as present-day Liberta, Bethesda, and Pares, located just outside Betty's Hope Plantation. By the 1860s, a good portion of the island's population was living in "free villages" (Dyde 2000:177; Lowes 1994a:46). By the 1850s, almost 6,000 freed Antiguans were still working on the island's estates for low wages, with others barely surviving on the margins of the old estates. The lack of effort toward developing any new agricultural crops or generating revenue to stimulate economic growth and improve the lives of Antiguans was further hampered by those dedicated to maintaining the status quo (Dyde 2000:168–170).

In the 1840s, the colonial office approved a plan to import indentured laborers. The first wave of Portuguese immigrants arrived in 1847 following a massive famine on the island. Between 1835 and 1882, some 2,000 Portuguese arrived from Madeira and a lesser number from Cape Verde (Lowes 1994a:43–44). They were supplemented by laborers coming from India, China, and the Middle East, who would later comprise Antigua's middle class (Williams 2007:25). Emancipation was also marked by violence. As Lightfoot (2015:6–7) explains, violence was a part of the post-Emancipation landscape as public disorderly conduct, as well as personal and collective violence played out on Antigua and elsewhere in the Caribbean, culminating with the labor riot of 1858. By the 1900s, a struggling sugar industry faced competing markets (Dyde 2000:185). Self-determination for the Antiguan people was a long way off, but their resilience persevered. As Samuel Smith expressed, "I hope that the day will never come again when our people have to suffer indignity like my generation and others have to. I am here to watch and see until the lord take me home" (Smith and Smith 1986:162).

Final Days

In the end, technology could not save Betty's Hope or similar plantations from their eventual demise. The sugar economy was dealt a major blow with the introduction of the sugar beet for widescale production in central Europe beginning in the early nineteenth century (Cooke and Scott 1993:xiv; Winner 1993:2–3). The opening up of the market to sugar from India, Mauritius, and Cuba also greatly depressed the British sugar market.

Technological innovations, however promising, could not guarantee economic stability for the long term. By 1895, sugar production had dropped by 60%, and Antigua was reduced to 78 plantations, with Betty's Hope as one of the last holdouts (Dyde 2000:203, 206–207). Before the plantation system completely collapsed, one last attempt was the implementation of a central sugar factory, the Antigua Sugar Factory, established in 1904, whose derelict remains can be seen from Factory Road as drivers pass by on their way into town and to the airport (figure A.1).

The factory was constructed on the old Gunthorpes estate and employed a number of islanders and linked the enterprise to the surviving sugar-producing estates like Betty's Hope through a railway system. After extended haggling, an agreement was finally reached between the Antigua Sugar Factory and William Gerald Codrington in 1936, when a contract was drawn up to devote 475 acres at Betty's Hope to cane production and to maintain a railway for transporting the cane stalks from Betty's Hope to the factory (figure A.2; NAAB Codrington Papers:C65).

By the mid-1930s, Betty's Hope was falling into disrepair. Tractors were introduced at Betty's Hope, but with the onset of the Great Depression and the resulting steep drop in sugar prices, they had little effect. Although

Above: Figure A.1. Old Sugar Factory, Factory Road, Antigua.

Left: Figure A.2. Restored locomotive from Betty's Hope.

Betty's Hope retained a significant labor force, wages for the island's laborers were significantly reduced (Lawrence 2015:177–178). After the sale of Betty's Hope in 1944 to Antigua Syndicate Estates, Ltd., sugarcane was still grown on the property until the Antigua Sugar Factory ceased operations in 1972.

During the final days of Betty's Hope, reminiscences of laborers were recorded. One of the most fascinating was Gigi Carr, who lived and worked at Betty's Hope. She recalled that local villagers from nearby Pares Village would do their laundry by taking water from one of the large four cisterns and then washing in the shade of nearby trees (Lawrence 2015:182). Carr also offered that life on the estate was no picnic: "You have to turn out to work every day, can't miss dat unless you really really sick and sometime they say really sick and sometimes the maninger would ride his harse and come by where you live and find out if you inside and hear if you answer and say you 'Yes Massa' . . . Everything 'yes and no Massa!' Can't just call dem by dere name, no sir . . ." (Lawrence 2015:180–181). Others, such as Harcourt Crump, remembered playing cricket on the lawn by the windmills (Lawrence 2015:188).

To some, the legacy of Betty's Hope is a painful reminder of a past that some would rather forget, whereas others consider it as part of a collective history. As visitors wander the grounds today, the full extent of the experiences of those who once lived and worked at Betty's Hope can never be fully understood or appreciated. Attempts at a historical archaeology of this landscape and others like it are a start, opening the door to a fuller understanding of the nature of slavery, the roots of capitalism, and the abiding passion for sugar that helped fuel these developments. It is up to the island nations of the Caribbean to reckon with their past in the ways they deem most appropriate. Historical archaeology can offer a way to access that past.

REFERENCES CITED

Unpublished Archival Records

ABBREVIATIONS

BLCP	British Library, London, Codrington Papers
NAAB	National Archives of Antigua and Barbuda, St. John, Antigua
NACO	National Archives, Kew, Richmond, Surrey (England), Colonial Office
SFU	Simon Fraser University Library, Burnaby, British Columbia
UCL	University College London

Baker, Robert

1749. *A New and Exact Map of the Island of Antigua in America According to an Actual and Accurate Survey Made in the Year 1746, 1747 and 1748.* Printed and sold by T. Bowles next ye Chapter House in St. Paul's Church-Yard, London. Original four sheets held at the John Carter Brown Library, Providence, Rhode Island. Electronic document, https://jcb.lunaimaging.com/luna/servlet/view/search/who/Baker%252C%2BRobert?q=Antigua+Baker&sort=normalized_date%2Cfile_name%2Csource_author%2Csource_title, accessed January 23, 2019.

British Library, London: Codrington Papers (BLCP)

Correspondence, Accounts, etc., Relating to the Codrington Family's Estates in Antigua and Barbuda (D1610/RP2616, 37 microfilm reels).

Codrington, Lady Georgiana Somerset

1843–1844. Diaries of Travels of Lady Georgiana Codrington in the West Indies. Gloucestershire Archives, England (D1610/F43).

Codrington, Sir William

Correspondence (C Series), Maps and Plans (P Series), and Estate Papers (E Series) etc. Relating to the Codrington Family's Estates in Antigua and Barbuda. Gloucestershire Archives, England (D1610).

1693–1747. Antigua Assembly Minute Books, Vols. 314–324.

1779–1782. Codrington Letterbook. James Marshall and Marie-Louise Osborn Collection, Beinecke Rare Book and Manuscript Library, Yale University, New Haven, Connecticut.

Fortescue, J. W. (editor)

1901 [1690]. America and West Indies: June 1690: Letter from Christopher Codrington to Lords of Trade and Plantations, June 4, 1690 (#927). In *Calendar of State Papers Colonial, America and West Indies: Volume 13, 1689–1692.* Her Majesty's Stationery Office, London. Electronic document, http://www.british-history.ac.uk/cal-state-papers/colonial/america-west-indies/vol13/pp276-291, accessed March 19, 2016.

Horneck, Kane William
1752. A Report of the State of the Fortifications in the Island of Antigua and Particularly Those of English Harbour in the Said Island in the year 1752. Antigua, West Indies. John Carter Brown Library, Annual Report, 1941, pp. 22–25.
Lancey, Thomas F.
1831. Antigua: Outline Map of the Island Showing Roads, Forts, Chapels, Main Settlements, etc. National Archives, Kew, Richmond, Surrey (England) (MR 1/1070).
Lovell, Langford
ca. 1790–1792. Letters to William Codrington (London: November 16, 1790 and January 18, 1792). Codrington Correspondence, 1743–1851. Microfilm. Kodak, Recordak Division, London.
Martin, Josiah
December 10, 1736. Letter to Robert Freeman (New York). Josiah Martin Letterbook, Martin Family Papers (Add MS 41346-41475). British Library, London.
National Archives, Kew, Richmond, Surrey (England), Colonial Office (NACO), Leeward Islands
Miscellanea. Naval Office Returns (Shipping Returns) for Antigua, 1704–1720; Nevis, 1683–1715; St. Christopher, 1683–1706; Montserrat, 1704–1715, 1784; Anguilla, 1787. Montserrat, Account of Produce, 1787 (CO 157/1).
Board of Trade. (CO 152/5:39, CO 152/7:L98, CO 152/16:S65 and 153:8. pp. 215–219).
National Archives, Kew, Richmond, Surrey (England), Colonial Office (NACO) and Predecessors
Acts of Antigua. 43 vols. (CO 8/1–4).
Antigua Sessional Papers. 81 vols. (CO 9/1–4 and 9/20).
Maps and Plans. Series 1, 2,975 flat sheets and vols. (CO 700).
National Archives of Antigua and Barbuda, St. John's, Antigua (NAAB)
1693–1747. Antigua Assembly Minute Books, Vols. 314–324.
1737. Antigua Council Minutes. Judges' Report, January 25, 1737, and Conspiracy Trials, November 1736–January 1737. PRO CO 9/10, fols. 49–91 (trials), 116, 119 (report); 9/11, fols. 35–44.
Codrington Papers. Correspondence (C Series), Maps and Plans (P Series), and Estate Papers (E Series) etc. Relating to the Codrington Family's Estates in Antigua and Barbuda. (D1610).
Redhead, Samuel
Letters from Samuel Redhead (Antigua) to William Codrington (London; June 8, 1754, May 3, 1755, July 18, 1758, December 12, 1762, March 21, 1771, March 6, 1772, and May 25, 1773). Codrington Correspondence, 1743–1851. Microfilm. Kodak, Recordak Division, London.
Reynolds, Dennis
Letter from Dennis Reynolds (Antigua) to William Codrington (London: May 22, 1792). Codrington Correspondence, 1743–1851. Microfilm. Kodak, Recordak Division, London.

Simon Fraser University Library, Burnaby, British Columbia (SFU)
1807. Sugar Accounts of C. B. Codrington, with M. Trattle, Merchant 1802–1807. Codrington Papers (microfilm 440, section 16, D1610 A60/1).
1813. Sugar Accounts of C. B. Codrington, with M. Trattle, Merchant 1807–1813. Codrington Papers (microfilm 441, section 1, D1610 A60/2).
1828. Account Sales Book of C. B. Codrington for Sugar and Wool 1824–1828. Codrington Papers (microfilm 441, section 2, D1610 A61).
1838. Annual Statements of Total Sugar Crop 1801–1838, with Calculations of Profits from Antigua and Barbuda 1707–1830, etc. Codrington Papers (microfilm 441, section 3, D1610 A62).
University College London (UCL)
2017. Legacies of British Slave-Ownership Database. Electronic document, https://www.ucl.ac.uk/lbs/claim/view, accessed July 27, 2017.

Secondary Sources

Abbott, George C.
1964. The West Indian Sugar Industry, with Some Long Term Projections of Supply to 1975. *Social and Economic Studies* 13(1):1–37.
Abbott, R. Tucker
1974. *American Seashells: The Marine Mollusca of the Atlantic and Pacific Coasts of North America.* 2nd ed. Van Nostrand Reinhold, New York.
Alexander, Edward P.
1997. *The Museum in America: Innovators and Pioneers.* Alta Mira Press, Walnut Creek, California.
2008. *Museums in Motion: An Introduction to the History and Functions of Museums.* Alta Mira Press, Walnut Creek, California.
Anderson, Jennifer L.
2012. *Mahogany: The Costs of Luxury in Early America.* Harvard University Press, Cambridge, Massachusetts.
Antigua Almanac and Register
1843. Antigua, St. Johns. Electronic document, http://www.antiguahistory.net/museumtest01/links-to-resources/, accessed January 23, 2019.
Arcangeli, Myriam
2015. *Sherds of History: Domestic Life in Colonial Guadeloupe.* University Press of Florida, Gainesville.
Armstrong, Douglas V.
1990. *The Old Village and the Great House: An Archaeological and Historical Examination of Drax Hall Plantation, St. Ann's Bay, Jamaica.* University of Illinois Press, Champaign.
1999. Archaeology and Ethnohistory of the Caribbean Plantation. In *"I, Too, Am America": Archaeological Studies of African-American Life,* edited by Theresa A. Singleton, pp. 173–192. University of Virginia Press, Charlottesville.
2003. *Creole Transformation from Slavery to Freedom: Historical Archaeology of the East End Community, St. John, Virgin Islands.* University Press of Florida, Gainesville.

Ashmore, Wendy

2018. Why the Archaeology of Political Ecology Matters. In *Uneven Terrain: Archaeologies of Political Ecology*, edited by J. K. Millhauser, C. T. Morehart, and S. Juarez, pp. 175–184. Archeological Papers of the American Anthropological Association 29:5–29. Wiley, New York.

Auchinleck, Gilbert G.

1956. *The Rainfall of Antigua and Barbuda: Compiled from Available Records.* Antigua Sugar Association, St. Johns, Antigua and Barbuda.

Bakker, Henk

1999. *Sugar Cane Cultivation and Management.* Springer Science and Business Media, New York.

Ball, Charles

1854. *Slavery in the United States. A Narrative of the Life and Adventures of Charles Ball, a Black Man, Who Lived Forty Years in Maryland, South Carolina, and Georgia, As a Slave.* 3rd ed. Western, Pittsburgh, Pennsylvania.

Barber, Sarah

2012. Who Owns Knowledge? Heritage, Intellectual Property and Access in and to the History of Antigua and Barbuda. *Archival Science* 12:1–17.

Barbot, John

1732. *A Description of the Coasts of North and South Guinea.* Churchill Brothers, London.

Baronetage of England, Ireland, Nova Scotia, Great Britain and the United Kingdom

2016. Electronic document, http://www.leighrayment.com/baronetage.htm, accessed March 23, 2016.

Bates, Lynsey A., John M. Chenoweth, and James A. Delle (editors)

2016. *Archaeologies of Slavery and Freedom in the Caribbean: Exploring the Spaces in Between.* University of Florida Press, Gainesville.

Bayly, Christopher A.

2004. *The Birth of the Modern World 1780–1914: Global Connections and Comparisons.* Blackwell, Malden, Massachusetts.

Beckles, Hilary McD.

1989. *Natural Rebels: A Social History of Enslaved Black Women in Barbados.* Rutgers University Press, New Brunswick, New Jersey.

1997. Capitalism, Slavery and Caribbean Modernity. *Callaloo* 20(4):777–789.

Beinart, William, and Lotte Hughs

2007. *Environment and Empire.* Oxford University Press, Oxford, England.

Bemis, Samuel Flagg

1962. American Foreign Policy and the Blessings of Liberty. *The American Historical Review* 67(2):291–305.

Bennet, J. Harry, Jr.

1958. *Bondsmen and Bishops: Slavery and Apprenticeships on the Codrington Plantations of Barbados, 1710–1838.* University of California Press, Berkeley.

Berland, Alexander J., Sarah E. Metcalfe, and Georgina H. Endfield

2013. Documentary-Derived Chronologies of Rainfall Variability in Antigua, Lesser Antilles, 1770–1890. *Climate of the Past* 9(3):1331–1343.

Berleant-Schiller, Riva
1977. Production and Division of Labor in a West Indian Peasant Community. *American Ethnologist* 4(2):253–272.
Berleant-Schiller, Riva, and Susan Lowes, with Milton Benjamin (compilers)
1995. Antigua and Barbuda: An Annotated Critical Bibliography. *World Bibliographic Series,* Vol. 182. Clio Press, Oxford, England.
Bickham, Troy
2008. Eating the Empire: Intersections of Food, Cookery and Imperialism in Eighteenth-Century Britain. *Past & Present* 198(1):71–109.
Blackburn, Robin
1997. *The Making of New World Slavery from the Baroque to the Modern 1492–1800.* Verso, London.
Blacklands Research and Extension Center
2014. *Environmental Policy Integrated Climate Model: User's Manual Version 0810.* Electronic document, http://epicapex.tamu.edu/wp-content/blogs.dir/482/files/2013/02/EPIC.0810-User-Manual.pdf, accessed July 27, 2017.
Boger, Rebecca, Sophia Perdikaris, Amy E. Potter, John Mussington, Reginald Murphy, Louise Thomas, Calvin Gore, and Dwight Finch
2014. Water Resources and the Historic Wells of Barbuda: Tradition, Heritage and Hope for a Sustainable Future. *Island Studies Journal* 9(2):327–342.
Bowen, Joanne
1975. Probate Inventories: An Evaluation from the Perspective of Zooarchaeology and Agricultural History at Mott Farm. *Historical Archaeology* 9:11–25.
1998. To Market, To Market: Animal Husbandry in New England. *Historical Archaeology* 32(3):137–152.
2000. *The Starving Time at Jamestown: Faunal Analysis of Pit 1, Pit 3, the Bulwark Ditch, Ditch 6, Ditch 7, and Midden 1.* Submitted to William M. Kelso, Jamestown Rediscovery, and Association for the Preservation of Virginia Antiquities. Report on file at the Jamestown Rediscovery Center, Virginia.
2006. Appendix I: Military Foodways at Fort Pelham, a Faunal Analysis. In *The Line of Forts: Historical Archaeology on the Colonial Frontier of Massachusetts,* edited by Michael D. Coe, pp. 147–168. University Press of New England, Lebanon, New Hampshire.
Bradley, Geoffrey
2008. *Barbados: The Sugar Story.* Herbert, Christ Church, Barbados.
Brown, Laurence
2002. Monuments to Freedom, Monuments to Nation: The Politics of Emancipation and Remembrance in the Eastern Caribbean. *Slavery and Abolition* 23(3):93–116.
Brown, Vincent
2008. *The Reaper's Garden: Death and Power in the World of Atlantic Slavery.* Harvard University Press, Cambridge, Massachusetts.
Buisseret, David
1973. The Elusive Deodand: A Study of the Fortified Refuges of the Lesser Antilles. *Journal of Caribbean History* 6–7:43–80.

1996. The Estate Map of the Old World. In *Rural Images: Estate Maps in the Old and New Worlds,* edited by David Buisseret, pp. 5–26. University of Chicago Press, Chicago.

2010. Charles Boucher of Jamaica and the Establishment of Greenwich Longitude. *Imago Mundi* 62(2):239–247.

Bush, Barbara

1990. *Slave Women in Caribbean Society 1650–1838.* Indiana University Press, Bloomington.

Campbell, J. C., John Radke, J. T. Gless, and R. M. Wirtshafter

1992. An Application of Linear Programming and Geographic Information Systems: Cropland Allocation in Antigua. *Environment and Planning A: Economy and Space* 24(4):535–549.

Carney, Judith A., and Richard Nicholas Rosomoff

2009. *In the Shadow of Slavery: Africa's Botanical Legacy in the Atlantic World.* University of California Press, Berkeley.

Carrington, Selwyn

2000. The State of the Debate on the Role of Capitalism in Ending the Slave System. In *Caribbean Slavery in the Atlantic World: A Student Reader,* edited by Verene Shepherd and Hilary McD. Beckles, pp. 1042–1054. Markus Wiener, Princeton, New Jersey.

Carstensen, Birgit

1993. *Betty's Hope: An Antiguan Sugar Plantation.* Betty's Hope Trust, St. John's, Antigua.

Chapman, Fredrik Henrik af

2006 [1768]. *Architectura Navalis Mercatoria: The Classic of 18th-Century Naval Architecture.* Dover, Mineola, New York.

Chase, Diane Z., and Arlen F. Chase

2014. Path Dependency in the Rise and Denouement of a Classic Maya City: The Case of Caracol, Belize. In *The Resilience and Vulnerability of Ancient Landscapes: Transforming Maya Archaeology through IHOPE,* edited by A. L. Chase and V. L. Scarborough, pp. 142–154. Archeological Papers of the American Anthropological Association, Number 24. Wiley, New York.

Claassen, Cheryl

1998. *Shells.* Cambridge University Press, Cambridge, England.

Cleland, Charles E.

1970. Comparison of the Faunal Remains from French and British Refuse Pits at Fort Michilimackinac: A Study in Changing Subsistence Patterns. *Canadian Historic Sites: Occasional Papers in Archaeology and History* 3:8–23.

Coe, Michael

2006. *The Line of Forts: Historical Archaeology on the Colonial Frontier of Massachusetts.* University Press of New England, Lebanon, New Hampshire.

Conn, Stephen

1998. *Museums and Intellectual Life, 1876–1926.* University of Chicago Press, Chicago.

Cooke, D. A., and R. K. Scott
1993. Introduction. In *The Sugar Beet Crop: Science into Practice*, edited by D. A Cooke and R. K. Scott, pp. xiv–xix. Chapman & Hall, London.
Crader, Diana C.
1984. The Zooarchaeology of the Storehouse and the Dry Well at Monticello. *American Antiquity* 49(3):542–558.
1990. Slave Diet at Monticello. *American Antiquity* 55(4):690–717.
Crist, Raymond E.
1954. Changing Cultural Landscapes in Antigua, BWI. *American Journal of Economics and Sociology* 13(3):225–232.
Curtin, Philip
1969. *The Atlantic Slave Trade: A Census*. University of Wisconsin Press, Madison.
Dana, John Cotton
1917a. *The Gloom of the Museum*. Elm Tree, Woodstock, New York.
1917b. *The New Museum*. Elm Tree, Woodstock, New York.
Darby, Megan
2018. Hurricane-Hit Barbudans Protest to Save Communal Land from Developers. Electronic document, http://www.climatechangenews.com/2018/04/19/hurricane-hit-barbudans-protest-save-communal-land-developers/, accessed July 25, 2018.
Dare, Paul M.
2005. Shadow Analysis in High-Resolution Satellite Imagery of Urban Areas. *Photogrammetric Engineering & Remote Sensing* 71(2):169–177.
Dator, James
2015. Frank Travels: Space, Power and Slave Mobility in the British Leeward Islands, c. 1700–1730. *Slavery & Abolition* 36(2):1–25.
Davis, N. Darnell
1887. *The Cavaliers & Roundheads of Barbados, 1650–1652: With Some Account of the Early History of Barbados*. Argosy, Georgetown, British Guiana.
Davy, John
1971 [1854]. *The West Indies Before and Since Slave Emancipation Comprising the Windward and Leeward Islands' Military Command Founded on Notes and Observations Collected During a Three Years' Residence*. Cash, London. 1971 facsimile ed. Frank Cass, London.
Dawdy, Shannon Lee
2008. *Building the Devil's Empire: French Colonial New Orleans*. University of Chicago Press, Chicago.
2010. "A Wild Taste": Food and Colonialism in Eighteenth-Century Louisiana. *Ethnohistory* 57(3):389–414.
De Reu, Jeroen, Gertjan Plets, Geert Verhoeven, Philippe De Smedt, Machteld Bats, Bart Cherretté, Wouter De Maeyer, Jasper Deconynck, Davy Herremans, Pieter Laloo, Marc Van Meirvenne, and Wim De Clercq
2013. Towards a Three-Dimensional Cost-Effective Registration of the Archaeological Heritage. *Journal of Archaeological Science* 40(2):1108–1121.

Deerr, Nöel
1950. *The History of Sugar*, Vol. II. Chapman and Hall, London.
deGrandpre, Andrew
2017. On Tiny Barbuda, a 300-Year-Old Civilization Has Been "Extinguished." *Washington Post*. September 15. Washington, D.C. Electronic document, https://www.washingtonpost.com/news/capital-weather-gang/wp/2017/09/15/on-tiny-barbuda-a-300-year-old-civilization-has-been-extinguished/?noredirect=on&utm_term=.7cec103312ca, accessed January 23, 2019.
Delle, James A.
2014. Archaeology and the "Tensions of Empire." In *Rethinking Colonial Pasts through Archaeology*, edited by N. Ferris, R. Harrison, and M. V. Wilcox, pp. 333–347. Oxford University Press, New York.
Descantes, Christophe, Robert J. Speakman, and Michael D. Glascock
2009. Neutron Activation Analysis of Ceramics from Five Archaeological Sites in Antigua, West Indies. *Bulletin of the Peabody Museum of Natural History* 50(1):147–155.
Deur, Douglas, Adam Dick, Kim Recalma-Clutesi, and Nancy Turner
2015. Kwakwaka'wakw "Clam Gardens." *Human Ecology* 43(2):201–212.
Devonian Foundation
1988. *Index to the Codrington Papers*. Devonian Foundation, Calgary, Canada.
Dietler, Michael
2010. *Archaeologies of Colonialism: Consumption, Entanglement, and Violence in Ancient Mediterranean France*. University of California Press, Berkeley.
Dillman, Jefferson
2015. *Colonizing Paradise: Landscape and Empire in the British West Indies*. University of Alabama Press, Tuscaloosa.
Douglas, Mary
1972. Deciphering a Meal. *Daedalus* 101(1):61–81.
Dunn, Richard S.
1972. *Sugar and Slaves: The Rise of the Planter Class in the English West Indies, 1624–1713*. University of North Carolina Press, Chapel Hill.
Durie, Alastair
1993. Imitation in Scottish Eighteenth-Century Textiles: The Drive to Establish the Manufacture of Osnaburg Linen. *Journal of Design History* 6(2):71–76.
Dyde, Brian
2000. *A History of Antigua: The Unsuspected Isle*. MacMillan Education, London.
Easterling, William E., Norman J. Rosenberg, Mary S. McKenney, C. Allen Jones, Paul T. Dyke, and J. R. Williams
1992. Preparing the Erosion Productivity Impact Calculator (EPIC) Model to Simulate Crop Response to Climate Change and the Direct Effects of CO_2. *Agricultural and Forest Meteorology* 59(1):17–34.
EDAX Smart Insight
2014. Elemental Analysis through a Plastic Barrier. Application Note—Micro-XRF. Electronic document, http://www.edax.com/download/Elemental_Analysis_Through_Plastic_Barrier_HR.pdf, accessed July 27, 2017.

Edelson, S. Max
2007. Clearing Swamps, Harvesting Forests: Trees and the Making of a Plantation Landscape in the Colonial South Carolina Low Country. *Agricultural History* 81(3):381–406.
Edwards, Bryan
1793. *The History, Civil and Commercial, of the British West Indies*, Vol. II. Luke White, Dublin.
Edwards, Megan E.
2011. Virginia Ham: The Local and Global of Colonial Foodways. *Food and Foodways* 19:56–73.
Elleman, Bruce A., and S. C. M. Paine (editors)
2013. *Commerce Raiding: Historical Case Studies, 1755–2009.* Naval War College Press, Newport, Rhode Island.
Eltis, David
2001. The Volume and Structure of the Transatlantic Slave Trade: A Reassessment. *William and Mary Quarterly* 58(1):17–46.
2006. The British Transatlantic Slave Trade before 1714: Annual Estimates and Direction. In *The Lesser Antilles in the Age of European Expansion*, edited by R. L. Paquette and S. L. Engerman, pp. 182–205. University Press of Florida, Gainesville.
Engerman, Stanley L.
1996. Europe, the Lesser Antilles, and Economic Expansion 1600–1800. In *The Lesser Antilles in the Age of European Expansion*, edited by R. L. Paquette and S. L. Engerman, pp. 147–165. University Press of Florida, Gainesville.
Equinao, Olaudah
2003. *Olaudah Equiano: The Interesting Narrative and Other Writings.* Edited by Vincent Carretta. Penguin, New York.
Espersen, Ryan
2017. "Better Than We": Landscapes and Materialities of Race, Class, and Gender in Pre-Emancipation Colonial Saba, Dutch Caribbean. PhD dissertation, Department of Archaeology, University of Leiden, Netherlands.
Espinosa, Mariola
2014. The Question of Racial Immunity to Yellow Fever in History and Historiography. *Social Science History* 38(3–4):437–453.
Evans, Chris
2012. The Plantation Hoe: The Rise and Fall of an Atlantic Commodity, 1650–1850. *William and Mary Quarterly* 69(1):71–100.
Ferguson, Jeffrey R., and Brandi MacDonald
2017. *Instrumental Neutron Activation Analysis of Historic Ceramics and Clays from the Betty's Hope Site, Antigua.* Report on file, University of Missouri, Columbia.
Fisher, Christopher T.
2005. Demographic and Landscape Change in the Lake Pátzcuaro Basin, Mexico: Abandoning the Garden. *American Anthropologist* 107(1):87–95.

Fonstad, Mark A., James T. Dietrich, Brittany C. Courville, Jennifer L. Jensen, and Patrice E. Carbonneau

2013. Topographic Structure from Motion: A New Development in Photogrammetric Measurement. *Earth Surface Processes and Landforms* 38(4):421–430.

Fox, Georgia L.

2014. Archaeological Investigations at Betty's Hope Plantation, Antigua: Some Preliminary Thoughts on Theory. In *Bitasion Archaeologie des Habitations—Plantations des Petites Antilles,* edited by K. Kelly and B. Berard, pp. 33–41. Sidestone Press, Leiden, Netherlands.

2015. *The Archaeology of Tobacco and Smoking.* University Press of Florida, Gainesville.

2016. Transformations, Economics, and Bitter Outcomes: Archaeological Investigations at Betty's Hope Plantation—A Case Study Reflecting 300 Years of Caribbean Sugar Production. *Economic Anthropology* 3(2):228–239.

Franklin, Adrian

1997. An Unpopular Food? The Distaste for Fish and the Decline of Fish Consumption in Britain. *Food and Foodways* 7(4):227–264.

Galloway, J. H.

1985. Tradition and Innovation in the American Sugar Industry, c. 1500–1800: An Explanation. *Annals of the Association of American Geographers* 75(3):334–351.

2005. *The Sugar Cane Industry: An Historical Geography from Its Origins to 1914.* Cambridge University Press, New York.

Garside, A. L., M. J. Bell, and R. C. Magarey

2001. Monoculture Yield Decline—Fact Not Fiction. *Proceedings of the International Society of Sugar Cane Technology* 24:16–21.

Gaspar, David Barry

1978. The Antigua Slave Conspiracy of 1736: A Case Study of the Origins of Collective Resistance. *William and Mary Quarterly* 35(2):308–323.

1985. *Bondmen and Rebels: A Study of Master-Slave Relations in Antigua, with Implications for Colonial British America.* Johns Hopkins University Press, Baltimore.

1988. Slavery, Amelioration, and Sunday Markets in Antigua, 1823–1831. *Slavery and Abolition* 9(1):1–28.

1992. Working the System: Antigua Slaves and Their Struggle to Live. *Slavery and Abolition* 13(3):131–135.

1993. Sugar Cultivation and Slave Life in Antigua before 1800. In *Cultivation and Culture: Labor and the Shaping of Slave Life in the Americas,* edited by I. Berlin and P. D. Morgan, pp. 101–123. University of Virginia Press, Charlottesville.

Gassman, Phillip W., Jimmy R. Williams, Verel W. Benson, R. C. Izaurralde, Larry M. Hauck, C. A. Jones, Jay D. Atwood, James R. Kiniry, and Joan D. Flowers

2005. *Historical Development and Applications of the EPIC and APEX Models.* Center for Agricultural and Rural Development, Iowa State University, Ames.

Genet, Nicole

1996. *La faïence de Place Royale.* Collection Patrimoines, série Dossiers, 45. Publications du Québec, Sainte-Foy, Québec.

George, Tim
2017. Barbudan Land Ownership: A 200 Year Old Freedom Put At Risk Following Hurricane Irma. Electronic document, https://www.opendemocracy.net/en/beyond-trafficking-and-slavery/barbudan-land-ownership-200-year-old-freedom-won-by-emancipated-slaves-and-, accessed September 22, 2017.
Giraldo, Alexander
2018. Religion and Resistance—Obeah: The Ultimate Resistance. Electronic document, https://scholar.library.miami.edu/slaves/Religion/religion.html, accessed January 23, 2019.
Given, M.
2004. *The Archaeology of the Colonized.* Routledge, London.
Godbout, Geneviève
2016. A House in Waiting: Food and Hospitality on Antiguan Plantations, 1783–1904. PhD dissertation, Department of Anthropology, University of Chicago, Chicago.
Gomez, Michael
1998. *Exchanging Our Country Marks: The Transformation of African Identities in the Colonial and Antebellum South.* University of North Carolina Press, Chapel Hill.
Gonzalez Scollard, Edith
2008. Raising Cane: Sugar, People and the Environment in Nineteenth-Century Antigua, West Indies. PhD dissertation, Department of Anthropology, The City University of New York, New York.
Goodwin, Conrad
1994. Betty's Hope Windmill: An Unexpected Problem. *Historical Archaeology* 28(1):99–110.
Goody, Jack
1982. *Cooking, Cuisine and Class: A Study in Comparative Sociology.* Cambridge University Press, Cambridge, England.
Gosden, Christopher
2005. *Archaeology and Colonialism: Cultural Contact from 5000 BC to the Present.* Cambridge University Press, Cambridge, England.
Gosden, Christopher, and Lesley Head
1994. Landscape—A Usefully Ambiguous Concept. *Archaeology in Oceania* 29:113–116.
Goudge, Charlotte
2017. The Trans-Atlantic Spirit of Change: The Effect of Social, Technological, and Environmental Change on Rum Production in the British Caribbean During the Late Eighteenth and Nineteenth Centuries. PhD dissertation, Department of Anthropology and Archaeology, University of Bristol, Bristol, England.
Gould, Kenneth A., and Tammy L. Lewis
2018. Green Gentrification and Disaster Capitalism in Barbuda. *NACLA Report on the Americas* 50(2):148–153.
Goveia, Elsa V.
1965. *Slave Society in the British Leeward Islands at the End of the Eighteenth Century.* Yale University Press, New Haven, Connecticut.

Håkansson, N. Thomas, and Mats Widgren
2014. *Landesque Capital: The Historical Ecology of Enduring Landscape Modifications.* Left Coast Press, Walnut Creek, California.
Hall, Catherine
2002. *Civilising Subjects: Metropole and Colony in the English Imagination 1830–1867.* University of Chicago Press, Chicago.
Hall, Douglas
2010. Incalculability as a Feature of Sugar Production during the Eighteenth Century. In *West Indian Business History: Enterprise and Entrepreneurship,* edited by B. W. Higman and K. E. A. Monteith, pp. 55–72. University of the West Indies Press, Kingston, Jamaica.
Hall, Gwendolyn
2005. *Slavery and African Ethnicities in the Americas: Restoring the Links.* University of North Carolina Press, Chapel Hill.
Handler, Jerome S.
1964. Notes on Pottery-Making in Antigua. *Man: A Record of Anthropological Science* 64:150–151.
2000. Slave Medicine and Obeah in Barbados, Circa 1650 to 1834. *New West Indian Guide* 74(1/2):57–90.
Handler, Jerome S., and JoAnn Jacoby
1996. Slave Names and Naming in Barbados, 1650–1830. *William and Mary Quarterly* 53(4):685–728.
Harlow, Vincent T.
1990. *Christopher Codrington, 1668–1710.* Reprinted. St. Martin's, New York. Original published 1928, Hurst, London.
Harris, David R.
1965. *Plants, Animals, and Man in the Outer Leeward Islands, West Indies: An Ecological Study of Antigua, Barbuda, and Anguilla.* University of California Press, Berkeley.
Hart, James L.
2013. Forts on the Frontier: Adapting European Military Engineering to North America. In *The Archaeology of French and Indian War Frontier Forts,* edited by Lawrence E. Babits and Stephanie Gandulla, pp. 17–51. University Press of Florida, Gainesville.
Hartman, Saidiya
2006. *Lose Your Mother: A Journey Along the Atlantic Slave Route.* Farrar, Straus and Giroux, New York.
Harvey, Karen
2012. Ritual Encounters: Punch Parties and Masculinity in the Eighteenth Century. *Past & Present* 214(1):165–203.
Harwin, Steve, and Arko Lucieer
2012. Assessing the Accuracy of Georeferenced Point Clouds Produced Via Multi-View Stereopsis from Unmanned Aerial Vehicle (UAV) Imagery. *Remote Sensing* 4(6):1573–1599.

Hauser, Mark W.
2008. *An Archaeology of Black Markets: Local Ceramics and Economies in Eighteenth-Century Jamaica.* University Press of Florida, Gainesville.
2011. Uneven Topographies: Archaeology of Plantations and Caribbean Slave Economies. In *The Archaeology of Capitalism in Colonial Contexts: Postcolonial Historical Archaeologies,* edited by S. K. Croucher and L. Weiss, pp. 121–142. Springer, New York.
Hauser, Mark W., and Jerome Handler
2009. Change in Small Scale Pottery Manufacture in Antigua, West Indies. *African Diaspora Archaeology Newsletter* 12(4):1–14.
Hauser, Mark W., and Dan Hicks
2007. Colonialism and Landscape: Power, Materiality and Scales of Analysis in Caribbean Historical Archaeology. In *Envisioning Landscape: Situations and Standpoints in Archaeology and Heritage,* edited by D. Hicks, L. McAtackney, and G. Fairclough, pp. 251–277. Left Coast Press, Walnut Creek, California.
Haynes, William M. (editor)
2014. *CRC Handbook of Chemistry and Physics,* 95th ed. CRC Press, Boca Raton, Florida.
Heath, Barbara
2016. Cowrie Shells, Global Trade, and Local Exchange: Piecing Together the Evidence for Colonial Virginia. *Historical Archaeology* 50(2):17–46.
Heath, B. J., and A. Bennett
2000. "The Little Spots Allow'd Them": The Archaeological Study of African-American Yards. *Historical Archaeology* 34(2):38–55.
Hein, George E.
1998. *Learning in the Museum.* Routledge, London.
Hicks, Dan
2007. *"The Garden of the World." An Historical Archaeology of Sugar Landscapes in the Eastern Caribbean.* BAR International Series 1632, Oxford, England.
Higham, C. S. S.
1926. The General Assembly of the Leeward Islands. *English Historical Review* 41(162):190–209.
Higman, Barry William
1984. *Slave Populations of the British Caribbean 1807–1834.* Johns Hopkins University Press, Baltimore.
1986. Plantation Maps as Sources for the Study of West Indian Ethnohistory. In *Ethnohistory: A Researcher's Guide,* edited by G. Wiedman, pp. 107–135. Department of Anthropology, College of William and Mary, Williamsburg, Virginia.
2000. Physical and Economic Environments. In *Caribbean Slavery in the Atlantic World: A Student Reader,* edited by Verene Shepherd and Hilary McD. Beckles, pp. 365–389. Ian Randle, Kingston, Jamaica.
2001. *Jamaica Surveyed: Plantation Maps and Plans of the Eighteenth and Nineteenth Centuries.* University Press of the West Indies, Kingston, Jamaica.

Higman, Barry William, George A. Aarons, Karlis Karklins, and Elizabeth Jean Reitz
1998. *Montpelier, Jamaica: A Plantation Community in Slavery and Freedom, 1739–1912.* University Press of the West Indies, Kingston, Jamaica.
Hobson, Daphne
2007. The Domestic Architecture of the Earliest British Colonies in the American Tropics: A Study of the Houses of the Caribbean "Leeward" Islands of St. Christopher, Nevis, Antigua and Montserrat, 1624–1726. PhD dissertation, Department of Architecture, Georgia Institute of Technology, Atlanta.
Hodder, Ian
2012. *Entangled: An Archaeology of the Relationships between Humans and Things.* Wiley-Blackwell, New York.
Holt, Thomas C.
1992. *The Problem of Freedom: Race, Labor, and Politics in Jamaica and Britain, 1832–1938.* Johns Hopkins University Press, Baltimore.
Hornborg, Alf, and Carole Crumley (editors)
2007. *The World System and the Earth System.* Left Coast Press, Walnut Creek, California.
Huang, M., J. Gallichand, T. Dang, and M. Shao
2006. An Evaluation of EPIC Soil Water and Yield Components in the Gully Region of Loess Plateau, China. *The Journal of Agricultural Science* 144(04):339–348.
International Council on Monuments and Sites (ICOMOS)
2015a. History. Electronic document, https://www.icomos.org/en/about-icomos/mission-and-vision/history, accessed May 5, 2015.
2015b. ICOMOS' Mission. Electronic document, https://www.icomos.org/en/about-icomos/mission-and-vision/icomos-mission, accessed March 1, 2015.
Jamaica Observer (Kingston, Jamaica)
2015. Indian Men Arrested for Practicing Obeah in Antigua. 11 November. Electronic document, http://www.jamaicaobserver.com/news/Indian-men-arrested-for-practicing-obeah-in-Antigua, accessed January 23, 2019.
Jarvis, Sondra A.
1997. When There's Nothing Better to Eat: Subsistence Strategies in Eighteenth Century Bermuda. Master's thesis, Department of Anthropology, College of William and Mary, Williamsburg, Virginia.
Jelks, Edward
2007. French Ceramics from the Fortress of Louisbourg, Nova Scotia. In *French Colonial Pottery. An International Conference,* edited by George Avery, pp. 97–150. Northwestern State University Press, Natchitoches, Louisiana.
Johnson, Sandy
2012a. The Gaming Pieces of Little Bay, Part I—Origins, Manufacture, and Arrival on Montserrat. *Little Bay Plantation Archaeology and Heritage Project, Montserrat, West Indies* (blog), April 20, 2012. http://littlebayarchaeology.blogspot.com/2012/03/introduction-gaming-pieces-of-carr.html, accessed January 23, 2019.
2012b. The Gaming Pieces of Little Bay, Part II—Social Context and Usage. *Little Bay Plantation Archaeology and Heritage Project, Montserrat, West Indies* (blog), April

20, 2012. http://littlebayarchaeology.blogspot.com/2012/03/gaming-pieces-of-little-bay-part.html, accessed January 23, 2019.
Kajuna, Silas T. A. R.
2001. *Millet: Post-Harvest Operations.* Food and Agriculture Organization of the United Nations, Rome.
Kea, Ray
2012. *A Cultural and Social History of Ghana from the Seventeenth to the Nineteenth Century: The Gold Coast in the Age of Trans-Atlantic Slave Trade.* 2 vols. Edwin Mellen Press, Lewiston, New York.
Keegan, William F., and Lisabeth A. Carlson
2008. *Talking Taíno: Essays on Caribbean Natural History from a Native Perspective.* University of Alabama Press, Tuscaloosa.
Kelly, Kenneth G., Mark W. Hauser, Christophe Descantes, and Michael D. Glascock
2008. Compositional Analysis of French Colonial Ceramics: Implications for Understanding Trade and Exchange. *Journal of Caribbean Archaeology* 8:85–108.
Kelly, Kenneth G., and Elhadj Ibrahima
2015. Employing Archaeology to (Dis)Entangle the Nineteenth-Century Illegal Slave Trade on the Rio Pongo, Guinea. *Atlantic Studies* 12(3):317–335.
Kelly, Kenneth G., and Diane Wallman
2014. Foodways of Enslaved Laborers on French West Indian Plantations (18th–19th Century). *Afriques. Débats, méthodes et terrains d'histoire.* Electronic document, http://afriques.revues.org/1608, accessed July 8, 2015.
Kiple, Kenneth F.
1984. *The Caribbean Slave: A Biological History.* Cambridge University Press, Cambridge, England.
Kiple, Kenneth F., and Virginia H. Kiple
1980. Deficiency Diseases in the Caribbean. *Journal of Interdisciplinary History* 11(2):197–215.
Kirby, Benjamin Crossley
2015. Can You Point Me to Your Nearest Clay Source Please?: An XRF Study of Barbadian Ceramics. Master's thesis, Department of Anthropology, College of William & Mary, Williamsburg, Virginia.
Kirch, Patrick V.
1997. Microcosmic Histories: Island Perspectives on "Global" Change. *American Anthropologist* 99(1):30–42.
Klein, Herbert S.
2010. *The Atlantic Slave Trade.* 2nd ed. Cambridge University Press, Cambridge, England.
Klein, Naomi
2007. *The Shock Doctrine: The Rise of Disaster Capitalism.* Metropolitan Books, New York.
Klippel, W. E.
2001. Sugar Monoculture, Bovid Skeletal Part Frequencies, and Stable Carbon Isotopes:

Interpreting Enslaved African Diet at Brimstone Hill, St. Kitts, West Indies. *Journal of Archaeological Science* 28:1191–1198.

Kopytoff, Igor

1986. The Cultural Biography of Things. In *The Social Life of Things: Commodities in Cultural Perspective,* edited by Arjun Appadurai, pp. 64–92. Cambridge University Press, Cambridge, England.

Kuhn, Thomas

1962. *The Structure of Scientific Revolutions.* University of Chicago Press, Chicago.

Kupperman, Karen Ordahl

1984. Fear of Hot Climates in the Anglo-American Colonial Experience. *William and Mary Quarterly* 41(2):213–240.

Lanaghan, Mrs.

1991 [1844]. *Antigua and the Antiguans: A Full Account of the Colony and Its Inhabitants from the Time of the Caribs to the Present Day, Interspersed with Anecdotes and Legends: Also, an Impartial View of Slavery and the Free Labour Systems: The Statistics of the Island, and Biographical Notices of the Principal Families,* Vol I. Saunders and Otley, London. 1991 facsimile ed. MacMillan Education, London.

Landon, David B.

2009. An Update on Zooarchaeology and Historical Archaeology: Progress and Prospects. In *International Handbook of Historical Archaeology,* edited by T. Majewski and D. Gaimster, pp. 77–104. Springer Science+Business Media, New York.

Law, Robin

1986. Dahomey and the Slave Trade: Reflections on the Historiography of the Rise of Dahomey. *The Journal of African History* 27(2):237–267.

2004. *Ouidah: The Social History of a West African Slaving "Port," 1727–1892.* James Currey, Oxford, England.

2005. Ethnicities of Enslaved Africans in the Diaspora: On the Meanings of "Mina" (Again). *History in Africa* 32:247–267.

Lawrence, Joy

2015. *Barbuda and Betty's Hope: The Codrington Connection.* Sugarmill Tales, St. John's, Antigua.

Lawrence, Joy, and Veronica Gordon

2013. *The Footprints of Parham: The History of a Small Antiguan Town and Its Influence.* Sugarmill Tales, St. John's, Antigua.

Leech, Roger

2013. "Within Musquett Shott of Black Rock"—Johnson's Fort and the Early Defenses of Nevis, West Indies. In *First Forts: Essays on the Archaeology of Proto-Colonial Fortification,* edited by Eric Klingelhofer, pp. 127–138. Brill, Boston.

Legislature of Antigua

1805. *The Laws of the Island of Antigua: Consisting of the Acts of the Leeward Islands, Commencing 8th November 1690, ending 21st April 1798; and the Acts of Antigua, Commencing 10th April 1668, ending 7th May 1804.* 2 vols. Samuel Bagster, London.

Lenik, Stephan

2005. Historical Archaeological Approaches to Tamarind Trees in the U.S. Virgin Islands. In *Proceedings of the Twenty-First Congress of the International Association for Caribbean Archaeology*, edited by B. Reid, H. Petitjean Roget, and A. Curet, pp. 31–40. University of the West Indies School of Continuing Studies, St. Augustine, Trinidad and Tobago.

Lightfoot, Natasha

2015. *Troubling Freedom: Antigua and the Aftermath of British Emancipation.* Duke University Press, Durham, North Carolina.

Ligon, Richard

1673. *A True & Exact History of the Island of Barbadoes.* 2nd ed. Peter Parker, London.

London Evening Post

November 29, 1746, to December 2, 1746; Issue 2976. In *Burney Database of Eighteenth and Nineteenth Century Newspapers.* Electronic document, https://www.bl.uk/collection-guides/burney-collection, accessed September 22, 2017.

Low, Theodore L.

1942. *The Museum as a Social Instrument.* Metropolitan Museum of Art, and the American Association of Museums, New York.

Lowe, Robson

1951. *The Codrington Correspondence, 1743–1851; Being a Study of a Recently Discovered Dossier of Letters from the West Indian Islands of Antigua and Barbuda Mostly Addressed to the Codringtons of Dodington, with Especial Reference to the History of Those Adventurous Times and the Hitherto Unrecorded Postal History of the Antiguan Mail.* Robson Lowe, London.

Lowenthal, David, and Colin Clarke

1977. Slave-Breeding in Barbuda: The Past of a Negro Myth. *Annals of the New York Academy of Sciences* 292(1):510–535.

1979. Common Lands, Common Aims: The Distinctive Barbudan Community. In *Peasants, Plantations, and Rural Communities in the Caribbean,* edited by Malcolm Cross and Arnaud Marks, pp. 142–59. Royal Institute of Anthropology and Linguistics, Leiden, Netherlands.

Lowes, Susan

1994a. The Peculiar Class: The Formation, Collapse, and Reformation of the Middle Class in Antigua, West Indies, 1834–1940. PhD dissertation, Department of Anthropology, Columbia University, New York.

1994b. *Sugar and Empire: The History of Antigua and Barbuda.* Electronic document, http://antiguahistory.net/uploads/3/0/5/2/3052847/lowes_part01.pdf, accessed January 23, 2019.

1995. "They Couldn't Mash Ants": The Decline of the White and Non-White Elites in Antigua, 1834–1900. In *Small Islands, Large Questions: Society, Culture and Resistance in the Post-Emancipation Caribbean,* edited by K. F. Olwig, pp. 31–52. Frank Cass, London.

Luffman, John

1789. A *Brief Account of the Island of Antigua: Together With the Customs and Manners of Its Inhabitants, As Well White As Black: As Also an Accurate Statement of the Food,*

Cloathing, Labor, and Punishment, of Slaves: In Letters to a Friend Written in the Years 1786, 1787, 1788. T. Cadell in the Strand, London.

Lyons, Claire L., and John K. Papadopoulos

2002. *The Archaeology of Colonialism.* Getty Research Institute, Los Angeles.

Machling, Tessa C. S.

2012. *The Fortifications of Nevis, West Indies, from the 17th Century to the Present Day: Protected Interests?* Archaeopress, Oxford, England.

McNeill, J. R.

2010. *Mosquito Empires: Ecology and War in the Greater Caribbean, 1620–1914.* Cambridge University Press, New York.

Mancini, Francesco, Marco Dubbini, Mario Gattelli, Francesco Stecchi, Stefano Fabbri, and Giovanni Gabbianelli

2013. Using Unmanned Aerial Vehicles (UAV) for High-Resolution Reconstruction of Topography: The Structure from Motion Approach on Coastal Environments. *Remote Sensing* 5(12):6880–6898.

Mardia, Kantilal V., John T. Kent, and John M. Bibby

1979. *Multivariate Analysis.* Academic Press, London.

Martin, Samuel

1750. *An Essay Upon Plantership, Humbly Inscrib'd to All the Planters of the British Sugar-Colonies in America. The Second Edition, Corrected and Enlarged.* T. Smith, Antigua.

1784. An Essay upon Plantership. In *Annals of Agriculture and Other Useful Arts,* Vol. 18, edited by A. Young, pp. 236–320. Bury St. Edmunds, Suffolk, England.

1785. *An Essay on Plantership.* Robert Mearns, Antigua.

Marx, Jenifer

1992. *Pirates and Privateers of the Caribbean.* Krieger, Malabar, Florida.

Marx, Karl

1979 [1859]. *Kritik der Politischen Ökonomie.* Dietz Verlag, Berlin.

Mason, Keith

2013. The Absentee Planter and the Key Slave: Privilege, Patriarchalism, and Exploitation in the Early Eighteenth-Century Caribbean. *William and Mary Quarterly* 70(1):79–102.

Menard, Russell R.

2007. Plantation Empire: How Sugar and Tobacco Planters Built Their Industries and Raised an Empire. *Agricultural History* 81(3):309–332.

Meniketti, Marco G.

2006. Sugar Mills, Technology, and Environmental Change: A Case Study of Colonial Agro-Industrial Development in the Caribbean. *Journal of the Society for Industrial Archaeology* 32(1):53–80.

2015. *Sugar Cane Capitalism and Environmental Transformation: An Archaeology of Colonial Nevis, West Indies.* University of Alabama Press, Tuscaloosa.

Mennell, Stephen

1996. *All Manners of Food: Eating and Taste in England and France from the Middle Ages to the Present.* University of Illinois Press, Champaign.

Métreau, Laetitia (editor)
2017. *Identifier la céramique au Québec*. CELAT Press, Québec.
Meyer, Jan H., Rianto Van Antwerpen, and Eddie Meyer
1996. A Review of Soil Degradation and Management Research under Intensive Sugarcane Cropping. *Proceedings of the South African Sugar Technology Association* 70:22–28.
Midgett, Douglas K.
1984. Distorted Development: The Resuscitation of the Antiguan Sugar Industry. *Studies in Comparative International Development* 19(2):33–58.
Mintz, Sidney W.
1974. *Caribbean Transformations*. Aldine, Chicago.
1985. *Sweetness and Power: The Place of Sugar in Modern History*. Viking Penguin, New York.
Mintz, Sidney W., and Douglas Hall
1960. *The Origins of the Jamaican Internal Marketing System*. Yale University Press, New Haven, Connecticut.
Mintz, Sidney W., and Richard Price
1992. *The Birth of African-American Culture: An Anthropological Perspective*. Beacon Press, Boston.
Morehart, Christopher T., John K. Millhauser, and Santiago Juarez
2018. Archaeologies of Political Ecology—Genealogies, Problems, and Orientations. In *Uneven Terrain: Archaeologies of Political Ecology*, edited by J. K. Millhauser, C. T. Morehart, and S. Juarez. Archeological Papers of the American Anthropological Association 29:5–29. Wiley, New York.
Morgan, Philip
1997. The Cultural Implications of the Atlantic Slave Trade: African Regional Origins, American Destinations and New World Developments. *Slavery & Abolition* 18(1):122–145.
Mulcahy, Matthew
2006. *Hurricanes and Society in the British Greater Caribbean, 1624–1783*. Johns Hopkins University Press, Baltimore.
Mullins, Michael
1995. *Africa in America: Slave Acculturation and Resistance in the American South and the British Caribbean, 1736–1831*. University of Illinois Press, Champaign.
Murphy, Arthur Reginald
1999. The Prehistory of Antigua, Ceramic Age: Subsistence, Settlement, Culture and Adaptation within an Insular Environment. PhD dissertation, Department of Anthropology, University of Calgary, Alberta, Canada.
Museum of Antigua and Barbuda
2005. Electronic document, http://antiguahistory.net/Museum/bettyshope.htm#Hist, accessed July 27, 2017.
Museum of Antigua and Barbuda and Dockyard Museum
2012. Potworks Reservoir. Electronic document, http://www.antiguanice.com/v2/client.php?id=446, accessed November 2, 2015.

Neiman, Fraser D.
2008. The Lost Worlds of Monticello: An Evolutionary Perspective. *Journal of Archaeological Research* 64(2):161–193.
2018. Historical Ecology of Jefferson's Monticello Plantation. Paper presented at Emerging Ecologies: Archaeologies of Slavery, Landscape, & Environmental Change Symposium, May 4, 2018, University of California, Santa Cruz.
Newsom, Lee A., and Elizabeth S. Wing
2004. *On Land and Sea: Native American Uses of Biological Resources in the West Indies.* University of Alabama Press, Tuscaloosa.
Nichols, John Gough (editor)
1867. *The Herald and Genealogist,* Vol. 4. J. G. Nichols and R. C. Nichols, London.
Nicholson, Desmond V.
1994. *Antigua and Barbuda Forts.* Museum of Antigua and Barbuda, St. John's, Antigua.
2001. *Heritage Landmarks of Antigua and Barbuda.* Museum of Antigua and Barbuda, St. John's, Antigua.
Nicholson, Desmond V., and Brian Dyde
2014. *The Place Names of Antigua and Barbuda.* D. V. N. Foundation, St. John's, Antigua.
Ohman, Alexis
2013. Zooarchaeological Analysis of the Assemblages from PEH-001. Report on file, College of William and Mary, Williamsburg, Virginia.
2014. Saltfish vs. Parrotfish: The Role of Fish and Mollusks in English Colonial Foodways at Betty's Hope Plantation, Antigua, West Indies. Unpublished Master's thesis, Department of Archaeology, Simon Fraser University, Burnaby, British Columbia, Canada.
Oldmixon, John
1708. *The British Empire in America: Containing the History of the Discovery, Settlement, Progress and Present State of All the British Colonies on the Continent and Islands of America,* Vol. II. John Nicholson, London.
Oliver, Vere Langford
1894–1899. *The History of the Island of Antigua, one of the Leeward Caribbees in the West Indies, From the First Settlement in 1635 to the Present Time.* 3 vols. Mitchell and Hughes, London.
O'Malley, Gregory
2014. *Final Passages: The Intercolonial Slave Trade of British America, 1619–1807.* University of North Carolina Press, Chapel Hill.
Orser, Charles E., Jr.
2010. Twenty-First-Century Historical Archaeology. *Journal of Archaeological Research* 18:111–150.
O'Shaughnessy, Andrew Jackson
1994. The Stamp Act Crisis in the British Caribbean. *William and Mary Quarterly* 51(2):203–226.
2015. *An Empire Divided: The American Revolution and the British Caribbean.* University of Pennsylvania Press, Philadelphia.

Paquette, Robert L., and Stanley L. Engerman
1996. Introduction. In *The Lesser Antilles in the Age of European Expansion*, edited by R. L. Paquette and S. L. Engerman, pp. 1–13. University Press of Florida, Gainesville.
Parker, Matthew
2011. *The Sugar Barons: Family, Corruption, Empire, and War in the West Indies*. Walker, New York.
Petley, Christer
2012. Gluttony, Excess, and the Fall of the Planter Class in the British Caribbean. *Atlantic Studies: Global Currents* 9(1):85–106.
Pettigrew, William A.
2007. Free to Enslave: Politics and the Escalation of Britain's Transatlantic Slave Trade, 1688–1714. *William and Mary Quarterly* 64(1):3–38.
2013. *Freedom's Debt: The Royal African Company and the Politics of the Atlantic Slave Trade, 1672–1752*. University of North Carolina Press, Chapel Hill.
Phillips, Fred
2013. *Commonwealth Caribbean Constitutional Law*. Routledge, New York.
Pickles, John
2012. *A History of Spaces: Cartographic Reason, Mapping, and the Geo-Coded World*. Routledge, New York.
Pinckard, George
1816. *Notes on the West Indies, Including Observations Relative to the Creoles and Slaves of the Western Colonies, including the Indians of South America*, Vol. I. Baldwin, Cradock, and Paternoster, London.
Pittman, Bonnie
1999. Muses, Museums, and Memories. *Daedalus* 128(3):1–31.
Porter, Dale H.
1970. *The Abolition of the Slave Trade in England, 1784–1807*. Archon Books, Hamden, Connecticut.
Potter, Amy E.
2015. The Commons as a Tourist Commodity: Mapping Memories and Changing Sense of Place on the Island of Barbuda. In *Social Memory and Heritage Tourism Methodologies*, edited by Stephen P. Hanna, Amy E. Potter, E. Arnold Modlin, Perry Carter, and David L. Butler, pp. 109–128. Routledge, New York.
Praetzellis, Adrian, Grace Ziesing, and Mary Praetzellis
1997. *Tales of the Vasco. Los Vaqueros Project Finn/Report #5*. Anthropological Studies Center, Sonoma State University Academic Foundation, Rohnert Park, California.
Pratt, Suzanna M.
2015. Landscape Legacies of Sugarcane Monoculture at Betty's Hope Plantation, Antigua, West Indies. Master's thesis, Department of Anthropology, University of South Florida, Tampa.
Price, Richard
1966. Caribbean Fishing and Fishermen: A Historical Sketch. *American Anthropologist* 68(6):1363–1383.

Pulsipher, Lydia Mihelic

1986. *Seventeenth Century Montserrat: An Environmental Impact Statement.* Historical Geography Research Series No. 17. GeoBooks, Norwich, England.

1990. They Have Saturdays and Sundays to Feed Themselves. *Expedition* 32(2):24.

1994. The Landscapes and Ideational Roles of Caribbean Slave Gardens. In *The Archaeology of Garden and Field,* edited by Naomi F. Miller and Kathryn L. Gleason, pp. 202–222. University of Pennsylvania Press, Philadelphia.

Radburn, Nicholas

2015. Guinea Factors, Slave Sales, and the Profits of the Transatlantic Slave Trade in Late Eighteenth-Century Jamaica: The Case of John Tailyour. *William and Mary Quarterly* 72 (2):243–286.

Ragatz, Lowell J.

1928. *Fall of the Planter Class in the British Caribbean, 1763–1833.* The Century, New York.

Ramos, Arthur

1941. Acculturation among the Brazilian Negroes. *The Journal of Negro History* 26(2):244–250.

Rebovich, Samantha Anne

2015. Resistance and Reforms: Landscapes at Greencastle Estate, Antigua. In *The Limits of Tyranny: Archaeological Perspectives on the Struggle Against New World Slavery,* edited by James A. Delle, pp. 65–92. University of Knoxville Press, Tennessee.

Rediker, Marcus

2004. *Villains of All Nations: Atlantic Pirates in the Golden Age.* Beacon, Boston.

2007. *The Slave Ship: A Human History.* Viking Penguin, New York.

Reeves, Dache M.

1936. Aerial Photography and Archaeology. *American Antiquity* 2(2):102–107.

Reilly, Matthew C.

2014. At the Margins of the Plantation: Alternative Modernities and an Archaeology of the "Poor Whites" of Barbados. PhD dissertation, Department of Anthropology, Syracuse University, New York.

Reitz, Elizabeth J.

1979. Spanish and British Subsistence Strategies at St. Augustine, Florida, and Frederica, Georgia, Between 1565 and 1783. PhD dissertation, Department of Anthropology, University of Florida, Gainesville.

1992. Vertebrate Remains from Highland House and Codrington Castle, Barbuda. Manuscript on file, Zooarchaeology Laboratory, University of Georgia, Athens.

Reitz, Elizabeth J., and Nicholas Honerkamp

1983. British Colonial Subsistence Strategy on the Southeastern Coastal Plain. *Historical Archaeology* 17(2):4–26.

Richards, John F.

2003. *The Unending Frontier: An Environmental History of the Modern World.* University of California Press, Berkeley.

Richardson, Bonham C.
1992. *The Caribbean in the Wider World, 1492–1992: A Regional Geography.* Cambridge University Press, New York.
2004. *Igniting the Caribbean's Past Fire in British West Indian History.* University of North Carolina Press, Chapel Hill.

Richardson, David
1987. The Slave Trade, Sugar, and British Economic Growth, 1748–1776. *Journal of Interdisciplinary History* 17(4):739–769.

Roberts, Justin
2013. *Slavery and the Enlightenment in the British Atlantic, 1750–1807.* Cambridge University Press, Cambridge, England.

Robertson, Robert
2003. The Edible West Indian "Whelk" *Cittarium pica* (Gastropoda: Trochidae): Natural History with New Observations. *Proceedings of the Academy of Natural Sciences of Philadelphia* 153:27–47.

Rouessac, Francis, and Annick Rouessac
2007. *Chemical Analysis: Modern Instrumentation Methods and Techniques.* John Wiley & Sons, Chichester, England.

Rucker, Walter
2015. *Gold Coast Diasporas: Identity, Culture, and Power.* Indiana University Press, Bloomington.

Ryzewski, Krysta, and John F. Cherry
2016.Surveying a Long-Term Settlement on Potato Hill, Montserrat. In *Archaeologies of Slavery and Freedom in the Caribbean: Exploring the Spaces in Between,* edited by L. A. Bates, J. M. Chenoweth, and J. A. Delle, pp. 152–179. University of Florida Press, Gainesville.

Sackett, James R.
1982. Approaches to Style in Lithic Archaeology. *Journal of Anthropological Archaeology* 1:59–112.

Samford, Patricia
1996. The Archaeology of African-American Slavery and Material Culture. *William and Mary Quarterly* 53(1):87–114.

Scarry, Margaret C., and Elizabeth J. Reitz
1990. Herbs, Fish, Scum, and Vermin: Subsistence Strategies in Sixteenth-Century Spanish Florida. In *Columbian Consequences, 2: Archaeological and Historical Perspectives in the Spanish Borderlands East,* edited by David Hurst Thomas, pp. 343–354. Smithsonian Institution Press, Washington, DC.

Scott, Elizabeth M.
2001. Food and Social Relations at Nina Plantation. *American Anthropologist* 103(3): 671–690.
2008. Who Ate What? Archaeological Food Remains and Cultural Diversity. In *Case Studies in Environmental Archaeology,* 2nd ed., edited by Elizabeth Reitz, C. Margaret Scarry, and Sylvia J. Scudder, pp. 357–374. Springer Science, New York.

Serrell, Beverly
1996. *Exhibit Labels: An Interpretive Approach.* Alta Mira Press, Walnut Creek, California.
Shackel, Paul A.
2014. Craft to Wage Labour: Agency and Resistance in American Historical Archaeology. In *Agency in Archaeology,* edited by Marcia-Anne Dobres and John E. Robb, pp. 232–246. Routledge, London.
Sharples, Jason T.
2012. Hearing Whispers, Casting Shadows: Jailhouse Conversation and the Production of Knowledge During the Antigua Slave Conspiracy Investigation of 1736. In *Buried Lives: Incarcerated in Early America,* edited by Richard Bell and Michele Lise Tarter, pp. 35–59. University of Georgia Press, Athens.
Sheridan, Richard B.
1960. Samuel Martin, Innovating Sugar Planter of Antigua 1750–1776. *Agricultural History* 34(3):126–139.
1961. The Rise of a Colonial Gentry: A Case Study of Antigua, 1730–1775. *The Economic History Review* 13(3):342–357.
1974. *Sugar and Slavery: An Economic History of the British West Indies, 1623–1775.* Baltimore: Johns Hopkins University Press.
1989. Changing Sugar Technology and the Labour Nexus in the British Caribbean, 1750–1900, With Special Reference to Barbados and Jamaica. *Nieuwe West-Indische Gids/New West Indian Guide* 63(1/2):59–93.
Shuler, Kristina A.
2011. Life and Death on a Barbadian Sugar Plantation: Historic and Bioarchaeological Views of Infection and Mortality at Newton Plantation. *International Journal of Osteoarchaeology* 21:66–81.
Sichler, Judith A.
2003. Historic Period Foodways in the Danish West Indies (1718–1917): The Zooarchaeological Evidence from Cinnamon Bay and the East End, St. John, Virgin Islands. Unpublished PhD dissertation, Department of Anthropology, University of Tennessee, Knoxville.
Simmons, Anne M.
2017. Plans to rebuild hurricane-ravaged Barbuda reignite a decades-old land dispute. *Los Angeles Times*, October 11. Electronic document, http://www.latimes.com/world/global-development/la-fg-barbuda-land-dispute-20171009-story.html, accessed July 25, 2018.
Singleton, Theresa A.
2001. Slavery and Spatial Dialectics on Cuban Coffee Plantations. *World Archaeology* 33(1):98–114.
2006. African Diaspora Archaeology in Dialogue. In *Afro-Atlantic Dialogues: Anthropology in the Diaspora,* edited by K. A. Yelvington, pp. 249–287. School of American Research Press, Santa Fe, New Mexico.
Sluyter, Andrew
2009. The Role of Black Barbudans in the Establishment of Open-Range Cattle Herd-

ing in the Colonial Caribbean and South Carolina. *Journal of Historical Geography* 35(2):330–349.
2012. *Black Ranching Frontiers: African Cattle Herders of the Atlantic World, 1500–1900.* Yale University Press, New Haven, Connecticut.

Smith, Frederick H.
2005. *Caribbean Rum: A Social and Economic History.* University Press of Florida, Gainesville.

Smith, Keithlyn B., and Fernando C. Smith
1986. *To Shoot Hard Labor: The Life and Times of Samuel Smith an Antiguan Workingman 1877–1982.* Edan's, Scarborough, Ontario.

Snavely, Noah, Steven M. Seitz, and Richard Szeliski
2008. Modeling the World from Internet Photo Collections. *International Journal of Computer Vision* 80(2):189–210.

Snelgrave, William
1734. *A New Account of Some Parts of Guinea and the Slave Trade.* Crown in Ludgate Street, London.

Southey, Thomas
1827. *Chronological History of the West Indies.* 3 vols. Longman, Rees, Orme, Brown, and Green, London.

Stelten, Ruud
2013. *Archaeological Excavations at Schotsenhoek Plantation, St. Eustatius, Caribbean Netherlands An Early- to Mid-Eighteenth-Century Slave Settlement at a Sugar Plantation on the Caribbean's "Historical Gem."* St. Eustatius Center for Archaeological Research, St. Eustatius, Caribbean Netherlands.

Stine, Linda France, Melanie A. Cabak, and Mark D. Groover
1996. Blue Beads as African-American Cultural Symbols. *Historical Archaeology* 30(3):49–75.

Stoler, Ann Laura, and Frederick Cooper
1997. Between Metropole and Colony: Rethinking a Research Agenda. In *Tensions of Empire: Colonial Cultures in a Bourgeois World*, edited by F. Cooper and A. L. Stoler, pp. 1–56. University of California Press, Berkeley.

Suliman, Adela
2018. Communal Land Ownership in Barbuda a Myth, Says Prime Minister. *Reuters.* April 23, London. Electronic document, https://www.reuters.com/article/us-britain-barbuda-land/communal-land-ownership-in-barbuda-a-myth-says-prime-minister-idUSKBN1HU2D5, accessed July 25, 2018.

Swanson, Drew A.
2009. Wormsloe's Belly: The History of a Southern Plantation Through Food. *Southern Cultures* 15(4):50–66.

Sweet, James
2013. Defying Social Death: The Multiple Configurations of African Slave Family in the Atlantic World. *William and Mary Quarterly* 70(2):251–272.

Taubes, Gary
2016. *The Case Against Sugar.* Knopf, New York.

Texas A&M AgriLife Research
2014. EPIC and APEX Models. Electronic document, http://epicapex.tamu.edu/epic, accessed July 27, 2017.
Thomas, Brian W.
1998. Power and Community: The Archaeology of Slavery at the Hermitage Plantation. *American Antiquity* 63(4):531–551.
Thomasson, David A.
1994. Montserrat Kitchen Gardens: Social Functions and Development Potential. *Caribbean Geography* 5(1):20.
Thorne, R. G. (editor)
1986. *The History of Parliament: The House of Commons, 1790–1820, 1: Introductory Survey*. Secker and Warburg, London.
Thornton, John
1998. *Africa and Africans in the Making of the Atlantic World, 1400–1800*. Cambridge University Press, Cambridge, England.
Tombs, Robert
2015. *The English and Their History*. Knopf, New York.
Tomich, Dale Wayne
2000. The Other Face of Slave Labor: Provision Grounds and Internal Marketing in Martinique. In *Caribbean Slave Society and Economy*, edited by Hilary Beckles and Verene Shepherd, pp. 743–757. Markus Wiener, Princeton, New Jersey.
Tricarico, Anthony R.
2019. Environmental Legacies of Pre-Columbian and Historic Land Use in Antigua, West Indies. PhD dissertation, Department of Anthropology, University of South Florida, Tampa.
Trouillot, Michel-Rolph
1992. The Caribbean Region: An Open Frontier in Anthropological Theory. *Annual Review of Anthropology* 21:19–42.
2004. *Global Transformations: Anthropology and the Modern World*. Palgrave MacMillan, New York.
Tuma, Michael W.
2006. Ethnoarchaeology of Subsistence Behaviors Within a Rural African American Community: Implications for Interpreting Vertebrate Faunal Data from Slave Quarters Areas of Antebellum Plantation Sites. *Historical Archaeology* 40(4):1–26.
Tweedy, Margaret T.
1981. A History of Barbuda under the Codringtons, 1738–1833. PhD dissertation, Department of History, University of Birmingham, Birmingham, England.
Twiss, Katheryn
2012. The Archaeology of Food and Social Diversity. *Journal of Archaeological Research* 20:357–395.
Twitty, Michael W.
2011. Gardens. In *World of a Slave. Encyclopedia of the Material Life of Slaves in the United States*, edited by Martha B. Katz-Hyman and Kym S. Rice, pp. 245–250. Greenwood, Santa Barbara, California.

UKTea & Infusions Association
2018. Tea Glossary and FAQs. Electronic document, https://www.tea.co.uk/tea-faqs, accessed July 27, 2017.
United Nations
2005. United Nations Convention to Combat Desertification Draft National Action Plan for Antigua and Barbuda. Electronic document, https://knowledge.unccd.int/sites/default/files/naps/antigua_and_barbuda-eng2005.pdf, accessed July 27, 2017.
University of Florida
1994. *Four Monuments of Antigua: Proposed to the Organization of the American States for the Selection of the Monuments of the Greater Caribbean; St. John's Cathedral, Shirley Heights, Nelson's Dockyard, Betty's Hope Estate*. University of Florida College of Architecture, Preservation Institute Caribbean, Gainesville.
Veit, Richard, and Paul W. Schopp
1999. Who's Been Drinking on the Railroad? Archaeological Excavations at the Central Railroad of New Jersey's Lakehurst Shops. *Northeast Historical Archaeology* 28: 21-40.
Verhoeven, Geert J. J., Jo Loenders, Frank Vermeulenand, and Roald Docter
2009. Helikite Aerial Photography—A Versatile Means of Unmanned, Radio Controlled, Low-Altitude Aerial Archaeology. *Archaeological Prospection* 16(2):125–138.
Verrand, Laurence
2004. Fortifications Militaires de Martinique 1635–1845. *Journal of Caribbean History Special Publication* 1:11–28.
Voyages
2013. Trans-Atlantic Slave Database. *Emory University.* http://www.slavevoyages.org/, accessed January 23, 2019.
2017a. Trans-Atlantic Slave Trade Database, Estimates from 1501 to 1866. *Emory University.* http://www.slavevoyages.org/estimates/saoL7PAG, accessed January 15, 2017.
2017b. Trans-Atlantic Slave Trade Database, List of Voyages from 1772 to 1866. *Emory University.* http://www.slavevoyages.org/voyages/qZsNDqzt, accessed January 15, 2017.
Wallman, Diane
2014. Slave Community Food Ways on a French Colonial Plantation: Zooarchaeology at Habitation Crève Coeur, Martinique. In *Bitasion: Archeologie des Habitations—Plantations des Petits Antilles,* edited by K. Kelly and B. Berard, pp. 45–68. Sidestone Press, Leiden, Netherlands.
Wallman, Diane, and Sandrine Grouard
2017. Enslaved Laborer and Sharecropper Fishing Practices in 18th–19th Century Martinique: A Zooarchaeological and Ethnozoohistorical Study. *Journal of Ethnobiology* 37(3):398–420.
Walvin, James
1997. *Fruits of Empire: Exotic Produce and British Taste, 1660–1800.* Macmillan, Basingstoke, England.

Ward, John R.
1978. The Profitability of Sugar Planting in the British West Indies, 1650–1834. *The Economic History Review* 31(2):197–213.
Waselkov, Gregory A., and John A. Walthall
2002. Faience Styles in French Colonial North America: A Revised Classification. *Historical Archaeology* 36(1):62–78.
Waters, Christopher K.
2017. When the King's Last Argument Is But a Whimper: Artillery Deployment in Antigua's Colonial Fortifications. *Proceedings of the Ticonderoga Seminar on 18th-Century Artillery,* pp. 120–131. Ticonderoga, New York.
2018. Putting Forts in Their Place: The Politics of Defense in Antigua, 1670–1785. PhD dissertation, Department of Anthropology, Syracuse University, Syracuse, New York.
2019. Indefensible Landscapes: Power Dynamics, Social Relations, and Antigua's Eighteenth-Century Fortifications. In *Power, Political Economy and Historical Landscapes of the Modern World: Interdisciplinary Perspectives,* edited by C. R. DeCorse. State University of New York Press, Albany.
Waters, Michael R.
1992. *Principles of Geoarchaeology.* University of Arizona Press, Tucson.
Waterston, William
1859. *A Manual of Commerce.* Oliver and Boyd, Edinburgh.
Watters, David R.
1997. Historical Documentation and Archaeological Investigation of Codrington Castle, Barbuda, West Indies. *Annals of Carnegie Museum* 66(3):229–288.
Watts, David
1987. *The West Indies: Patterns of Development, Culture, and Environmental Change since 1492.* Cambridge University Press, Cambridge, England.
WeatherSpark
2014. Average Weather for Saint John's, Antigua and Barbuda. Electronic document, https://weatherspark.com/averages/33713/Saint-John-s-Antigua-Antigua-And-Barbuda-Antigua-and-Barbuda, accessed July 27, 2017.
Weaver, David B.
1988. The Evolution of a "Plantation" Tourism Landscape on the Caribbean Island of Antigua. *Tijdschrift voor economische en sociale geografie* 79(5):319–331.
Webb, Steven Saunders
2013. *Marlborough's America.* Yale University Press, New Haven, Connecticut.
Weigand, Phil C., Garman Harbottle, and Edward V. Sayre
1977. Turquoise Sources and Source Analysis: Mesoamerica and the Southwestern U.S.A. In *Exchange Systems in Prehistory,* edited by T. K. Earle and J. E. Ericson, pp. 15–34. Academic Press, New York.
Weil, Stephen E.
1990. Rethinking the Museum. In *Reinventing the Museum: Historical and Contemporary Perspectives on the Paradigm Shift,* edited by Gail Anderson, pp. 74–89. Alta Mira Press, Walnut Creek, California.

2002. *Making Museums Matter.* Smithsonian Institution Press, Washington, DC.
Wells, E. Christian, Karla L. Davis-Salazar, and David D. Kuehn
2013. Soilscape Legacies: Historical and Emerging Consequences of Socioecological Interactions in Honduras. In *Soils, Climate and Society: Archaeological Investigations in Ancient America,* edited by J. D. Wingard and S. E. Hayes, pp. 21–59. University Press of Colorado, Boulder.
Wells, E. Christian, Charlotte Goudge, Anthony R. Tricarico, Reginald Murphy, and Georgia L. Fox
2018a. Social and Environmental Impacts of British Colonial Rum Production at Betty's Hope Plantation, Antigua. In *Archaeologies of the British in Latin America,* edited by C. Orser Jr., pp. 235–253. Springer, New York.
Wells, E. Christian, Suzanna M. Pratt, Georgia L. Fox, Peter E. Siegel, Nicholas P. Dunning, and Reginald Murphy
2018b. Plantation Soilscapes: Initial and Cumulative Impacts of Colonial Agriculture in Antigua, West Indies. *Environmental Archaeology* 23(1):23–35.
Wells, E. Christian, Christopher K. Waters, Anthony R. Tricarico, and Georgia L. Fox
2017. Agroindustrial Soilscapes in the Caribbean: A Geochemical Perspective from Betty's Hope, Antigua. *Environmental Archaeology* 22(4):381–393.
Wells, Robert V.
1975. *The Population of the British Colonies in America before 1776: A Survey of Census Data.* Princeton University Press, Princeton, New Jersey.
Westoby, Matthew J., James Brasington, Neil F. Glasser, Michael J. Hambrey, and John M. Reynolds
2012. Structure-from-Motion Photogrammetry: A Low-Cost, Effective Tool for Geoscience Applications. *Geomorphology* 179:300–314.
Wilkie, Laurie A.
2004. Considering the Future of African American Archaeology. *Historical Archaeology* 38(1):109–123.
2016. Conclusion: Minding the Gaps in the Diaspora Web. In *Archaeologies of Slavery and Freedom in the Caribbean: Exploring the Spaces in Between,* edited by L. A. Bates, J. M. Chenoweth, and J. A. Delle, pp. 329–351. University of Florida Press, Gainesville.
Williams, Christolyn A.
2007. Labor Organization, Political Leadership, and Gender Exclusion in Antigua and Barbuda, 1917–1970. PhD dissertation, Department of Anthropology, City University of New York, New York.
Williams, Eric
1994 [1944]. *Capitalism & Slavery.* University of North Carolina Press, Chapel Hill.
Williams, Jimmy R.
1990. The Erosion-Productivity Impact Calculator (EPIC) Model: A Case History. *Philosophical Transactions of the Royal Society of London. Series B: Biological Sciences* 329(1255):421–428.
Williams, Jimmy R., C. Allan Jones, and P. T. Dyke
1984. The EPIC Model and Its Application. In *Proceedings of the International Sympo-*

sium on Minimum Data Sets for Agrotechnology Transfer, edited by Vrinda Kumble, pp. 111–121. International Crops Research Institute for the Semi-Arid Tropics (ICRSAT) Center, Patancheru, Andhra Pradesh, India.

Wilson, Anne C.

1973. *Food and Drink in Britain: From the Stone Age to Recent Times.* Penguin Books Ltd., Middlesex, England.

Wilson, Richard, and Alan Mackley

2000. *Creating Paradise: The Building of the English Country House, 1660–1880.* Hambledon Continuum, London.

Wingard, John D.

2013. Tilling the Fields and Building the Temples: Assessing the Relationship among Land, Labor, and Classic Maya Elite Power in the Copán Valley, Honduras. In *Soils, Climate and Society: Archaeological Investigations in Ancient America,* edited by John D. Wingard and Sue E. Hayes, pp. 131–155. University Press of Colorado, Boulder.

Wingard, John D., and Sue E. Hayes (editors)

2013. *Soils, Climate and Society: Archaeological Investigations in Ancient America.* University Press of Colorado, Boulder.

Winner, C.

1993. History of the Crop. In *The Sugar Beet Crop: Science into Practice,* edited by D. A. Cooke and R. K. Scott, pp. 1–35. Chapman & Hall, London.

Wood, Denis

1993. The Fine Line Between Mapping and Map Making. *Cartographica: The International Journal for Geographic Information and Geovisualization* 30(4):50–60.

Wood, Peter

1975. *Black Majority: Negroes in Colonial South Carolina from 1670 through the Stono Rebellion.* Norton, New York.

Wray, Leonard

1848. *The Practical Sugar Planter: A Complete Account of the Cultivation and Manufacture of the Sugar-Cane, According to the Latest and Most Improved Processes. Describing and Comparing the Different Systems Pursued in the East and West Indies and the Straits of Malacca.* Smith, Elder, London.

Zacek, Natalie A.

2007. Cultivating Virtue: Samuel Martin and the Paternal Ideal in the Eighteenth-Century English West Indies. *Wadabagei: A Journal of the Caribbean and Its Diasporas* 10(3):8–31.

2010. *Settler Society in the English Leeward Islands, 1670–1776.* Cambridge University Press, Cambridge, England.

Zahedieh, Nuala

1986. Trade, Plunder, and Economic Development in Early English Jamaica, 1655–89. *The Economic History Review, New Series* 39(2):205–222.

CONTRIBUTORS

Jennifer L. Anderson is associate professor of history in the Department of History at SUNY Stony Brook. She is the author of *Mahogany: The Costs of Luxury in Early America.*

Matthew Brown is assistant professor of archaeology in the Department of Sociology and Anthropology at Farmingdale State College, SUNY.

James F. Dator is assistant professor of history and Africana studies at Goucher College.

Erin Friedman is a PhD candidate at the Graduate Center CUNY in Earth and environmental sciences and senior fellow with the Urban Resilience to Extremes (UREx) network.

Georgia L. Fox is chair and professor of the Department of Anthropology at California State University, Chico.

Geneviève Godbout is associate member of the CÉLAT Graduate Research Centre at the Université Laval in Québec, Canada.

Charlotte Goudge received her PhD in archaeology and anthropology from the University of Bristol in 2017.

Benjamin C. Kirby is the manager of the archaeology lab at James Madison's Montpelier.

Amanda Kramp is assistant curator and exhibitions manager at the Turtle Bay Exploration Park's museum.

Cory Look is lecturer in anthropology and archaeology at Brooklyn College CUNY and serves as a researcher at the Anthropological Research Council.

Reginald Murphy is director of heritage resources at Nelson's Dockyard National Park and serves as secretary general, National Commission for UNESCO, Antigua and Barbuda.

Alexis Ohman is a zooarchaeologist and a PhD candidate at the College of William and Mary in Williamsburg, Virginia.

Suzanna M. Pratt serves as research analyst for the Joint Legislative Audit and Review Committee of the Washington State Legislature.

Anthony R. Tricarico is currently a doctoral student in applied anthropology at the University of South Florida, where he is conducting his dissertation research on the long-term impacts of agriculture on plantation landscapes in Antigua.

Christopher K. Waters is serving as a post-doctoral fellow at the National Parks Authority of Antigua and Barbuda.

E. Christian Wells is professor of anthropology and director of the Center for Brownfields Research and Redevelopment at the University of South Florida, where he has served previously as the founding director of the Office of Sustainability and deputy director of the Patel School of Global Sustainability.

INDEX

Page numbers in italics refer to illustrations.

RIPLEY P. BULLEN SERIES

Florida Museum of Natural History

Tacachale: Essays on the Indians of Florida and Southeastern Georgia during the Historic Period, edited by Jerald T. Milanich and Samuel Proctor (1978)

Aboriginal Subsistence Technology on the Southeastern Coastal Plain during the Late Prehistoric Period, by Lewis H. Larson (1980)

Cemochechobee: Archaeology of a Mississippian Ceremonial Center on the Chattahoochee River, by Frank T. Schnell, Vernon J. Knight Jr., and Gail S. Schnell (1981)

Fort Center: An Archaeological Site in the Lake Okeechobee Basin, by William H. Sears, with contributions by Elsie O'R. Sears and Karl T. Steinen (1982)

Perspectives on Gulf Coast Prehistory, edited by Dave D. Davis (1984)

Archaeology of Aboriginal Culture Change in the Interior Southeast: Depopulation during the Early Historic Period, by Marvin T. Smith (1987)

Apalachee: The Land between the Rivers, by John H. Hann (1988)

Key Marco's Buried Treasure: Archaeology and Adventure in the Nineteenth Century, by Marion Spjut Gilliland (1989)

First Encounters: Spanish Explorations in the Caribbean and the United States, 1492–1570, edited by Jerald T. Milanich and Susan Milbrath (1989)

Missions to the Calusa, edited and translated by John H. Hann, with an introduction by William H. Marquardt (1991)

Excavations on the Franciscan Frontier: Archaeology at the Fig Springs Mission, by Brent Richards Weisman (1992)

The People Who Discovered Columbus: The Prehistory of the Bahamas, by William F. Keegan (1992)

Hernando de Soto and the Indians of Florida, by Jerald T. Milanich and Charles Hudson (1993)

Foraging and Farming in the Eastern Woodlands, edited by C. Margaret Scarry (1993)

Puerto Real: The Archaeology of a Sixteenth-Century Spanish Town in Hispaniola, edited by Kathleen Deagan (1995)

Political Structure and Change in the Prehistoric Southeastern United States, edited by John F. Scarry (1996)

Bioarchaeology of Native American Adaptation in the Spanish Borderlands, edited by Brenda J. Baker and Lisa Kealhofer (1996)

A History of the Timucua Indians and Missions, by John H. Hann (1996)

Archaeology of the Mid-Holocene Southeast, edited by Kenneth E. Sassaman and David G. Anderson (1996)

The Indigenous People of the Caribbean, edited by Samuel M. Wilson (1997; first paperback edition, 1999)

Hernando de Soto among the Apalachee: The Archaeology of the First Winter Encampment, by Charles R. Ewen and John H. Hann (1998)

The Timucuan Chiefdoms of Spanish Florida, by John E. Worth: vol. 1, *Assimilation*; vol. 2, *Resistance and Destruction* (1998)

Ancient Earthen Enclosures of the Eastern Woodlands, edited by Robert C. Mainfort Jr. and Lynne P. Sullivan (1998)

An Environmental History of Northeast Florida, by James J. Miller (1998)

Precolumbian Architecture in Eastern North America, by William N. Morgan (1999)

Archaeology of Colonial Pensacola, edited by Judith A. Bense (1999)

Grit-Tempered: Early Women Archaeologists in the Southeastern United States, edited by Nancy Marie White, Lynne P. Sullivan, and Rochelle A. Marrinan (1999; first paperback edition, 2000)

Coosa: The Rise and Fall of a Southeastern Mississippian Chiefdom, by Marvin T. Smith (2000)

Religion, Power, and Politics in Colonial St. Augustine, by Robert L. Kapitzke (2001)
Bioarchaeology of Spanish Florida: The Impact of Colonialism, edited by Clark Spencer Larsen (2001)
Archaeological Studies of Gender in the Southeastern United States, edited by Jane M. Eastman and Christopher B. Rodning (2001)
The Archaeology of Traditions: Agency and History Before and After Columbus, edited by Timothy R. Pauketat (2001)
Foraging, Farming, and Coastal Biocultural Adaptation in Late Prehistoric North Carolina, by Dale L. Hutchinson (2002)
Windover: Multidisciplinary Investigations of an Early Archaic Florida Cemetery, edited by Glen H. Doran (2002)
Archaeology of the Everglades, by John W. Griffin (2002; first paperback edition, 2017)
Pioneer in Space and Time: John Mann Goggin and the Development of Florida Archaeology, by Brent Richards Weisman (2002)
Indians of Central and South Florida, 1513–1763, by John H. Hann (2003)
Presidio Santa María de Galve: A Struggle for Survival in Colonial Spanish Pensacola, edited by Judith A. Bense (2003)
Bioarchaeology of the Florida Gulf Coast: Adaptation, Conflict, and Change, by Dale L. Hutchinson (2004; first paperback edition, 2020)
The Myth of Syphilis: The Natural History of Treponematosis in North America, edited by Mary Lucas Powell and Della Collins Cook (2005)
The Florida Journals of Frank Hamilton Cushing, edited by Phyllis E. Kolianos and Brent R. Weisman (2005)
The Lost Florida Manuscript of Frank Hamilton Cushing, edited by Phyllis E. Kolianos and Brent R. Weisman (2005)
The Native American World Beyond Apalachee: West Florida and the Chattahoochee Valley, by John H. Hann (2006)
Tatham Mound and the Bioarchaeology of European Contact: Disease and Depopulation in Central Gulf Coast Florida, by Dale L. Hutchinson (2006)
Taíno Indian Myth and Practice: The Arrival of the Stranger King, by William F. Keegan (2007)
An Archaeology of Black Markets: Local Ceramics and Economies in Eighteenth-Century Jamaica, by Mark W. Hauser (2008; first paperback edition, 2013)
Mississippian Mortuary Practices: Beyond Hierarchy and the Representationist Perspective, edited by Lynne P. Sullivan and Robert C. Mainfort Jr. (2010; first paperback edition, 2012)
Bioarchaeology of Ethnogenesis in the Colonial Southeast, by Christopher M. Stojanowski (2010; first paperback edition, 2013)
French Colonial Archaeology in the Southeast and Caribbean, edited by Kenneth G. Kelly and Meredith D. Hardy (2011; first paperback edition, 2015)
Late Prehistoric Florida: Archaeology at the Edge of the Mississippian World, edited by Keith Ashley and Nancy Marie White (2012; first paperback edition, 2015)
Early and Middle Woodland Landscapes of the Southeast, edited by Alice P. Wright and Edward R. Henry (2013; first paperback edition, 2019)
Trends and Traditions in Southeastern Zooarchaeology, edited by Tanya M. Peres (2014)
New Histories of Pre-Columbian Florida, edited by Neill J. Wallis and Asa R. Randall (2014; first paperback edition, 2016)
Discovering Florida: First-Contact Narratives from Spanish Expeditions along the Lower Gulf Coast, edited and translated by John E. Worth (2014; first paperback edition, 2016)
Constructing Histories: Archaic Freshwater Shell Mounds and Social Landscapes of the St. Johns River, Florida, by Asa R. Randall (2015)
Archaeology of Early Colonial Interaction at El Chorro de Maíta, Cuba, by Roberto Valcárcel Rojas (2016)

Fort San Juan and the Limits of Empire: Colonialism and Household Practice at the Berry Site, edited by Robin A. Beck, Christopher B. Rodning, and David G. Moore (2016)

Rethinking Moundville and Its Hinterland, edited by Vincas P. Steponaitis and C. Margaret Scarry (2016; first paperback edition, 2019)

Handbook of Ceramic Animal Symbols in the Ancient Lesser Antilles, by Lawrence Waldron (2016)

Paleoindian Societies of the Coastal Southeast, by James S. Dunbar (2016; first paperback edition, 2019)

Gathering at Silver Glen: Community and History in Late Archaic Florida, by Zackary I. Gilmore (2016)

Cuban Archaeology in the Caribbean, edited by Ivan Roksandic (2016)

Archaeologies of Slavery and Freedom in the Caribbean: Exploring the Spaces in Between, edited by Lynsey A. Bates, John M. Chenoweth, and James A. Delle (2016; first paperback edition, 2018)

Setting the Table: Ceramics, Dining, and Cultural Exchange in Andalucía and La Florida, by Kathryn L. Ness (2017)

Simplicity, Equality, and Slavery: An Archaeology of Quakerism in the British Virgin Islands, 1740–1780, by John M. Chenoweth (2017)

Fit for War: Sustenance and Order in the Mid-Eighteenth-Century Catawba Nation, by Mary Elizabeth Fitts (2017)

Water from Stone: Archaeology and Conservation at Florida's Springs, by Jason O'Donoughue (2017)

Mississippian Beginnings, edited by Gregory D. Wilson (2017; first paperback edition, 2019)

Honoring Ancestors in Sacred Space: The Archaeology of an Eighteenth-Century African-Bahamian Cemetery, by Grace Turner (2017)

Investigating the Ordinary: Everyday Matters in Southeast Archaeology, edited by Sarah E. Price and Philip J. Carr (2018)

Harney Flats: A Florida Paleoindian Site, by I. Randolph Daniel Jr. and Michael Wisenbaker (2017)

Early Human Life on the Southeastern Coastal Plain, edited by Albert C. Goodyear and Christopher R. Moore (2018)

New Histories of Village Life at Crystal River, by Thomas J. Pluckhahn and Victor D. Thompson (2018)

The Archaeology of Villages in Eastern North America, edited by Jennifer Birch and Victor D. Thompson (2018)

The Cumberland River Archaic of Middle Tennessee, edited by Tanya M. Peres and Aaron Deter-Wolf (2019)

Pre-Columbian Art of the Caribbean, by Lawrence Waldron (2019)

Iconography and Wetsite Archaeology of Florida's Watery Realms, edited by Ryan Wheeler and Joanna Ostapkowicz (2019)

New Directions in the Search for the First Floridians, edited by David K. Thulman and Ervan G. Garrison (2019)

Archaeology of Domestic Landscapes of the Enslaved in the Caribbean, edited by James A. Delle and Elizabeth C. Clay (2019)

Cahokia in Context: Hegemony and Diaspora, edited by Charles H. McNutt and Ryan M. Parish (2020)

Bears: Archaeological and Ethnohistorical Perspectives in Native Eastern North America, edited by Heather A. Lapham and Gregory A. Waselkov (2020)

Contact, Colonialism, and Native Communities in the Southeastern United States, edited by Edmond A. Boudreaux III, Maureen Meyers, and Jay K. Johnson (2020)

An Archaeology and History of a Caribbean Sugar Plantation on Antigua, edited by Georgia L. Fox (2020)

CPSIA information can be obtained
at www.ICGtesting.com
Printed in the USA
JSHW020658211219
2957JS00001B/2

9 781683 401285